180 Standards-Based Problems of the Day GRADE 3

This book belongs to:

_____

180 Standards-Based Problems of the Day
GRADE 3

Dr. Bonita Manning-White
and Christine King

Copyright © 2023 by CKingEducation

All rights reserved.

No part of this book may be reproduced, distributed, or transmitted in any form or by any means, including photocopying, recording, or other electronic or mechanical methods, without the prior written permission of the author or publisher, except in the case of brief quotations embodied in critical reviews and certain other noncommercial uses permitted by copyright law.

Although the authors and publisher have made every effort to ensure that the information in this book was correct at press time, no liability is assumed for errors or omissions, or for losses or damages due to the information provided.

Published by CKingEducation

To contact CKingEducation or the authors about speaking, workshops, or ordering books in bulk, visit www.ckingeducation.com.

ISBN: 979-8-9890007-3-9

Lead writer: Dr. Bonita-Manning White
Editor: Ashley Baboolal
Series editor and book design: Christine King

Printed in the United States of America

Contents

Welcome to 3rd Grade and Goal Setting	15
POD #1: The Cookout	17
POD #2: A Bag of Apples	18
POD #3: Saturday Afternoon Fun	19
POD #4: Mail Delivery	20
POD #5: Number Riddle	21
POD #6: Ruby's Bookshelves	22
POD #7: Baseball Cards	23
POD #8: Summer Camp	24
POD #9: Vanilla Milkshakes	25
POD #10: State Fair	26
POD #11: Picking Strawberries	27
POD #12: Toy Trucks	28
POD #13: Family Vacation	29
POD #14: Roblox Points	30
POD #15: The Librarian	31
POD #16: Birthday Gift	32
POD #17: Jelly Beans	33
POD #18: The Bakery	34

POD #19: TikTok Videos — 35

POD #20: A Box of Pencils — 36

POD #21: Boy's Hockey Team Tryouts — 37

POD #22: Number Riddle — 38

POD #23: How Healthy Is Shrimp? — 39

POD #24: Bears — 40

POD #25: MineCraft — 41

POD #26: The Little Mermaid — 42

POD #27: Millipede Legs — 43

POD #28: Calorie Count — 44

POD #29: Walmart Sale — 45

POD #30: Collecting Box Tops — 46

POD #31: Treasure Box — 47

POD #32: Jordan's Book — 48

POD #33: The Parade — 49

POD #34: The Fundraiser — 50

POD #35: Chores — 51

POD #36: Lion King — 52

POD #37: BiC Gelocity Pens — 53

POD #38: Colored Pencils	54
POD #39: Summer Vacation	55
POD #40: Alice's Favorite Hobby	56
POD #41: STEAM Camp	57
POD #42: McDonald's	58
POD #43: Apple Watch	59
POD #44: Beehive	60
POD #45: Dave & Buster's	61
POD #46: Little Debbie Cakes	62
POD #47: World's Largest Edible Mushroom	63
POD #48: Agree or Disagree	64
POD #49: Pencils	65
POD #50: Flower Delivery	66
POD #51: Bookshelf	67
POD #52: Cars, Cars, Cars	68
POD #53: Lining Up	69
POD #54: Kelly's Secret Number	70
POD #55: Traveling Ants	71
POD #56: Tennis Balls	72

POD #57: Target					73

POD #58: Sharpening Pencils			74

POD #59: Friendship Bracelets			75

POD #60: Pumpkin Seeds				76

POD #61: Class Garden				77

POD #62: School Carnival			78

POD #63: Bottles of Water			79

POD #64: Bags of Apples				80

POD #65: State Fair				81

POD #66: Chewing Gum				82

POD #67: The Baker				83

POD #68: Recycling				84

POD #69: Math Problems				85

POD #70: Moving Day				86

POD #71: Pokémon Cards				87

POD #72: Chores					88

POD #73: Sharing				89

POD #74: School's Track Team			90

POD #75: Prize Box				91

POD #76: Cereal Bars	92
POD #77: Chairs	93
POD #78: The Farmer	94
POD #79: Geese Crossing the Road	95
POD #80: Pet Store	96
POD #81: Juice Boxes	97
POD #82: Box of Chocolates	98
POD #83: Field Trip	99
POD #84: Stickers	100
POD #85: Science Experiment	101
POD #86: The Bookshelf	102
POD #87: Bags of Candy	103
POD #88: Marching Band	104
POD #89: Marbles	105
POD #90: Dog Treats	106
POD #91: Sharing Cookies	107
POD #92: Bead Necklaces	108
POD #93: Counting Pennies	109
POD #94: New Pencils	110

POD #95: Arranging Stickers	111
POD #96: Fundraiser	112
POD #97: The Librarian	113
POD #98: Friday Treat	114
POD #99: New Game	115
POD #100: Square Tiles	116
POD #101: Maya's Rock Collection	117
POD #102: Marching Band	118
POD #103: Alex's Toy Cars	119
POD #104: Paper Butterflies	120
POD #105: Winter Musical	121
POD #106: My Story	122
POD #107: Gumballs	123
POD #108: Barbie Sticker Album	124
POD #109: Sleepover	125
POD #110: Bake Sale	126
POD #111: The Zookeeper	127
POD #112: A Game of Soccer	128
POD #113: The Science Museum	129

POD #114: Bags of Candy	130
POD #115: Paper Clips	131
POD #116: The Holidays	132
POD #117: Hockey Team Line	133
POD #118: Omelets	134
POD #119: Math Team	135
POD #120: Summer Job	136
POD #121: National Puppy Day Sale	137
POD #122: Fun Park	138
POD #123: Rolls of Tape	139
POD #124: Pencil Cap Eraser Sale	140
POD #125: Family Reunion T-Shirts	141
POD #126: Friendship Necklace Sale	142
POD #127: Cupcakes	143
POD #128: Peaches Galore	144
POD #129: The Glass Pitcher	145
POD #130: Fishing Trip	146
POD #131: Cross Stitching Pictures	147
POD #132: Halves or Not?	148

POD #133: Equal Parts	149
POD #134: Partitioning Rectangles	150
POD #135: Representing Fractional Amounts	151
POD #136: Brownies with Nuts	152
POD #137: Pepperoni Pizza	153
POD #138: Representing Thirds	154
POD #139: Dividing Paper	155
POD #140: Richard's Homework	156
POD #141: Peach Pie for Dessert	157
POD #142: Completing Homework	158
POD #143: Catching the School Bus	159
POD #144: Which is Larger?	160
POD #145: Which is Less?	161
POD #146: New York City Public School's Population	162
POD #147: Althea's Garden	163
POD #148: Movie Snacks	164
POD #149: Ordering Fractions	165
POD #150: Greater Than One-half	166
POD #151: Equal to One Whole	167

POD #152: Close to 0 — 168

POD #153: Going in Circles — 169

POD #154: Greater than 1 — 170

POD #155: Listing Fractions — 171

POD #156: Center Time — 172

POD #157: Which Fraction Doesn't Belong? — 173

POD #158: Subway Sandwich for Lunch — 174

POD #159: Lines, Lines, Lines — 175

POD #160: Folding Paper — 176

POD #161: Justine's Thinking — 177

POD #162: Torn Index Cards — 178

POD #163: Not Equivalent — 179

POD #164: Pie Eating Contest — 180

POD #165: Making Pizza — 181

POD #166: How Are They Similar? — 182

POD #167: Yani's Pie — 183

POD #168: Finding Equivalent Fractions — 184

POD #169: Time to Finish Homework — 185

POD #170: Math Class — 186

POD #171: Movie Time	187
POD #172: Bedtime	188
POD #173: Wrestling Event	189
POD #174: Basketball Practice	190
POD #175: Polygons	191
POD #176: Describing Quadrilaterals	192
POD #177: Guess My Quadrilateral	193
POD #178: Making Polygons	194
POD #179: Sorting Quadrilaterals	195
POD #180: Draw It	196
20 BONUS POD Templates	197
My Math Affirmations	217
3rd Grade Reflection	218

Welcome to 3rd Grade!

Did you know that you are a problem solver?
You are! We all are! That's what humans do - we solve problems!
This book is filled with 180 visualizable math word problems for you to:

3rd Grade Goal Setting

In order to achieve a desired result it helps to set goals.
Take a moment to write a friendly letter to your future self explaining how you feel about math and set some goals that you want to achieve.

Date: _____

_____,

_____

_____

_____

_____

_____

_____

_____

_____

_____

_____

_____

_____

Today's Affirmation: I am a mathematical thinker.

Name: _____   Date: _____   Time: ____:____

Problem of the Day #1: The Cookout

Maya's family is having a cookout. They are going to grill 22 hot dogs and 18 hamburgers. How many hotdogs and hamburgers will they grill?

Show Your Work:

Explain Your Thinking:

_____

_____

_____

_____

Base Ten Blocks, Ten Frames, 100-Bead Rekenrek, 100-Bead Number Line, Meter Stick, Measuring Tape, Number Line

Today's Affirmation: I am brave and I take risks in math.

Name: _____ Date: _____ Time: ___:____

Problem of the Day #2: A Bag of Apples

Mrs. Maxwell put 50 apples in a bag. She put some red and green apples in the bag. How many red apples can be in the bag? How many green apples can be in the bag?

Show Your Work:

Explain Your Thinking:

_____

_____

_____

_____

Base Ten Blocks, Ten Frames, 100-Bead Rekenrek, 100-Bead Number Line, Meter Stick, Measuring Tape, Number Line

Today's Affirmation: Math is more than just getting an answer.

Name: _____ Date: _____ Time: ____:____

Problem of the Day #3: Saturday Afternoon Fun

Tyrone spent 2 hours on Saturday afternoon playing his PlayStation 4. He played 2 rounds of Sonic Superstars. Tyrone scored a total of 90 points during round 1 and round 2. How many points could Tyrone have scored during round 1? How many points could he have scored during round 2?

Show Your Work:

Explain Your Thinking:

_____

_____

_____

_____

Base Ten Blocks, Ten Frames, 100-Bead Rekenrek, 100-Bead Number Line, Meter Stick, Measuring Tape, Number Line

Today's Affirmation: When I do math, I feel like a superhero.

Name: _____ Date: _____ Time: ____:____

Problem of the Day #4: Mail Delivery

On Tuesday, the mail carrier delivered 100 pieces of mail. He delivered some mail in the morning and some mail in the afternoon. How many pieces of mail could have been delivered in the morning and delivered in the afternoon?

Show Your Work:

Explain Your Thinking:

_____

_____

_____

_____

Base Ten Blocks, Ten Frames, 100-Bead Rekenrek, 100-Bead Number Line, Meter Stick, Measuring Tape, Number Line

Today's Affirmation: It is important that I explain and justify my thinking.

Name: _____ Date: _____ Time: ____:____

Problem of the Day #5: Number Riddle

Kerri saw a number riddle. It said:
- ❏ I am an odd number.
- ❏ I am greater than 200.
- ❏ The sum of my digits is 10.
- ❏ I am about 30 less than 300.

What number am I?

Show Your Work:

Explain Your Thinking:

_____

_____

_____

_____

Base Ten Blocks, Number Line, Number Grid

Today's Affirmation: We are all smart in different ways.

Name: _____ Date: _____ Time: ____:____

Problem of the Day #6: Ruby's Bookshelves

Ruby's mom bought her new bookshelves for all of her books. On one bookshelf, there are 136 nonfiction books. On the other bookshelf, there are 199 fiction books.

What is the Question?: _____

_____

Show Your Work:

Explain Your Thinking:

_____
_____
_____
_____

Base Ten Blocks, Ten Frames, Number Line, Number Grid, Number Bond

Today's Affirmation: Math helps me make sense of my world.

Name: _____ Date: _____ Time: ___:___

Problem of the Day #7: Baseball Cards

Dwayne and his brother Richard each have a collection of baseball cards. Dwayne has 390 baseball cards. Richard has 210 baseball cards. Richard gave all of his baseball cards to Dwayne as a gift. If Dwayne wanted to sell all of the baseball cards, how many baseball cards would he have to sell?

Show Your Work:

Explain Your Thinking:

_____

_____

_____

_____

Base Ten Blocks, Open Number Line, Number Grid, Number Bond, Bar Model/Tape Diagram

Today's Affirmation: Acting out math problems helps me understand what I am doing.

Name: _____ Date: _____ Time: ____:____

Problem of the Day #8: Summer Camp

Every year Kymriah goes to Ocean Breeze Summer Camp. This year there are 285 girls and 395 boys registered for camp. How many children attended Ocean Breeze Summer Camp this summer?

Show Your Work:

Explain Your Thinking:

_____

_____

_____

_____

Base Ten Blocks, Open Number Line, Number Grid, Number Bond, Bar Model/Tape Diagram

Today's Affirmation: Sketching out my thinking helps me see a problem more clearly.

Name: _____ Date: _____ Time: ____:____

Problem of the Day #9: Vanilla Milkshakes

McDonald's sold 513 vanilla milkshakes on Monday. On Tuesday, McDonald's sold 447 vanilla milkshakes. On Wednesday, McDonald's sold 353 vanilla milkshakes.

What is the Question?: _____

_____

Show Your Work:

Explain Your Thinking:

_____

_____

_____

_____

Base Ten Blocks, Open Number Line, Number Grid, Number Bond, Bar Model/Tape Diagram

Today's Affirmation: Making mistakes is how I learn new things.

Name: _____ Date: _____ Time: ____:____

Problem of the Day #10: State Fair

Three hundred twenty-two adults and five hundred seventy-eight children attended the State Fair on Tuesday night. What is the total number of people who attended the State Fair on Tuesday night?

Show Your Work:

Explain Your Thinking:

_____

_____

_____

_____

Base Ten Blocks, Open Number Line, Number Grid, Number Bond, Bar Model/Tape Diagram

Today's Affirmation: I can help others by asking questions.

Name: _____ Date: _____ Time: ____:____

Problem of the Day #11: Picking Strawberries

Amari and her sister went to the strawberry farm to pick strawberries. Amari picked 32 strawberries in the morning and 18 strawberries in the afternoon. Her sister picked 14 strawberries in the morning and 16 strawberries in the afternoon. They put all of their strawberries in one large basket. How many strawberries did they put in the basket?

Show Your Work:

Explain Your Thinking:

_____

_____

_____

_____

Base Ten Blocks, Unifix/Snap Cubes, 100-Bead Rekenrek, 100-Bead Number Line, Measuring Tape, Number Line

Today's Affirmation: I can do harder math problems by acting out the story.

Name: _____ Date: _____ Time: ____:____

Problem of the Day #12: Toy Trucks

Jesse has 3 toy trucks. There are 2 toy people in each truck. His friend Ryan also has 3 toy trucks. Ryan has 4 toy people in each truck. How many toy people are in the trucks?

Show Your Work:

Explain Your Thinking:

_____

_____

_____

_____

Equal Groups Model, Number Line, Unifix/Snap Cubes

Today's Affirmation: I am a problem solver.

Name: _____ Date: _____ Time: ____:____

Problem of the Day #13: Family Vacation

Martin's family spent a week driving to their favorite vacation spots. They drove 123 miles on Monday and 97 miles on Tuesday to get to Disney World. Then on Friday, they drove 136 miles to get to Legoland.

What is the Question?: _____

_____

Show Your Work:

Explain Your Thinking:

_____

_____

_____

_____

Base Ten Blocks, Open Number Line, Number Grid, Bar Model/Tape Diagram

Today's Affirmation: If I messed up yesterday, today is a new day, and I can try again.

Name: _____ Date: _____ Time: ____:____

Problem of the Day #14: Roblox Points

Rosie had 221 Roblox points in her account. For her birthday, she received 80 Roblox points, but used 1 point. Then, for making the Honor Roll she received 130 Roblox points. How many Roblox points does Rosie have now?

Show Your Work:

Explain Your Thinking:

_____

_____

_____

_____

Base Ten Blocks, Open Number Line, Number Grid, Bar Model/Tape Diagram

Today's Affirmation: Math helps me develop grit and perseverance.

Name: _____   Date: _____   Time: ____:____

Problem of the Day #15: The Librarian

The librarian received two boxes of books for the library. The first box had 100 books. The second box had 60 fewer books than the first box. What is the total number of books the librarian received?

Show Your Work:

Explain Your Thinking:

_____

_____

_____

_____

Base Ten Blocks, Open Number Line, Number Grid, Bar Model/Tape Diagram

Today's Affirmation: I don't need to be fast at math. I need to understand and that takes time.

Name: _____ Date: _____ Time: ___:____

Problem of the Day #16: Birthday Gift

Imani wants to buy her sister a birthday gift for $30. She earned $5 each day for 4 days. Her father gave her $5 for doing her chores. How much more money does Imani need in order to buy her sister a gift?

Show Your Work:

Explain Your Thinking:

_____

_____

_____

_____

Base Ten Blocks, Unifix/Snap Cubes, Equal Groups Model, Money (bills)

Today's Affirmation: Think positive thoughts. Negative thoughts don't help us learn and grow.

Name: _____ Date: _____ Time: ____:____

Problem of the Day #17: Jelly Beans

Malik went to the candy store to buy jelly beans. He put 30 strawberry jelly beans in a plastic bag. He also put some green apple jelly beans in the bag. Malik now has 92 jelly beans. How many green apple jelly beans did he put in the bag?

Show Your Work:

Explain Your Thinking:

_____

_____

_____

_____

Base Ten Blocks, Open Number Line, Number Grid, Bar Model/Tape Diagram, 100-Bead Number Line

Today's Affirmation: Mathematics makes me smarter because it makes me think!

Name: _____ Date: _____ Time: ____:____

Problem of the Day #18: The Bakery

The bakery made 2 dozen vanilla cupcakes and 3 dozen chocolate cupcakes to sell. At the end of the day, there were 18 cupcakes left. How many cupcakes did they sell?

Show Your Work:

Explain Your Thinking:

_____

_____

_____

_____

12 Count Egg Carton, Ruler, Base Ten Blocks, Bar Model/Tape Diagram, Equal Groups

Today's Affirmation: I am human and humans are mathematical beings.

Name: _____ Date: _____ Time: ____:____

Problem of the Day #19: TikTok Videos

Simone uploaded 48 videos to TikTok in July. In August, she uploaded 19 videos on TikTok. How many more videos did she upload in July than in August?

Show Your Work:

Explain Your Thinking:

_____

_____

_____

_____

Bar Model/Tape Diagram, Ten-Frame, Base Ten Blocks, Calendar, Hundreds Grid, Meter Stick

Today's Affirmation: Even when we know the answer, it can be fun to try out a different way.

Name: _____ Date: _____ Time: ___:___

Problem of the Day #20: A Box of Pencils

Malcolm's mom bought him a new box of pencils for school. During the first semester, Malcolm used 95 of his pencils. Now, Malcolm has 45 pencils. How many pencils were in the box to start?

Show Your Work:

Explain Your Thinking:

_____

_____

_____

_____

Bar Model/Tape Diagram, Base Ten Blocks, Number Grid

Today's Affirmation: I have a powerful mind!

Name: _____ Date: _____ Time: ___:___

Problem of the Day #21: Boy's Hockey Team Tryouts

There were 75 boys that tried out for the school's hockey team on Monday. On Tuesday, 55 boys tried out for the hockey team. Only 20 boys can make the hockey team.

What is the Question?: _____

_____

Show Your Work:

Explain Your Thinking:

_____

_____

_____

_____

Bar Model/Tape Diagram, Base Ten Blocks, Number Grid

Today's Affirmation: I am brilliant, bright, and getting better every day!

Name: _____   Date: _____   Time: ___:___

Problem of the Day #22: Number Riddle

Leslie heard a number riddle in school that she wanted her grandpa to solve.
- ❏ I am an odd number.
- ❏ I am about 40.
- ❏ I am the difference of 95 - 58.
- ❏ The sum of my digits is 10.

What number do you think her grandpa came up with? Why?

Show Your Work:

Explain Your Thinking:

_____

_____

_____

_____

Number Grid, Meter Stick, Measuring Tape, Number Line, Number Cards

Today's Affirmation: Answers are important, but questioning and solving are just as important.

Name: _____ Date: _____ Time: ___:___

Problem of the Day #23: How Healthy Is Shrimp?

Truth and her family went to dinner. Truth's mom was looking at the calories of two dishes and noticed that grilled shrimp had 79 calories and fried shrimp had 300 calories. How many more calories does the fried shrimp have than the grilled shrimp?

Show Your Work:

Explain Your Thinking:

_____

_____

_____

_____

Base Ten Blocks, Open Number Line, Number Bond, Number Grid, 100-Bead Number Line

Today's Affirmation: New solutions happen when we try new ways of doing things.

Name: _____ Date: _____ Time: ___:___

Problem of the Day #24: Bears

A mother bear and her 2 cubs were seen wandering through a neighborhood in the middle of the day. A bear biologist was called in to help capture the bears. The biologist confirmed that the weights of all the bears combined was about 400 pounds. What could be the weight of each bear?

Show Your Work:

Explain Your Thinking:

_____
_____
_____
_____

Base Ten Blocks, Hundred Number Grid

Today's Affirmation: By asking questions, I learn more and help others learn as well.

Name: _____ Date: _____ Time: ____:____

Problem of the Day #25: MineCraft

There were 479 players signed in to Minecraft on Saturday morning. How many more players can join before they reach the capacity of 1,000 players?

Show Your Work:

Explain Your Thinking:

_____

_____

_____

_____

Open Number Line, Number Grid, Bar Model/Tape Diagram, Base Ten Blocks

Today's Affirmation: I love learning! I am a learner.

Name: _____   Date: _____   Time: ____:____

Problem of the Day #26: The Little Mermaid

On Friday afternoon, 930 children's movie tickets were sold for "The Little Mermaid". There were 350 adult movie tickets sold for the same showing of "The Little Mermaid". How many more children's tickets were sold than adult tickets?

Show Your Work:

Explain Your Thinking:

_____

_____

_____

_____

Bar Model/Tape Diagram, Base Ten Blocks, Number Grid, Open Number Line

Name: _____ Date: _____ Time: ___:____

Problem of the Day #27: Millipede Legs

Millipede means "1,000 legs." Real millipedes only have about 750 legs. About how many fewer legs does a real millipede have than what the name means?

Show Your Work:

Explain Your Thinking:

_____

_____

_____

_____

Bar Model/Tape Diagram, Base Ten Blocks, Number Grid, Open Number Line

Today's Affirmation: I can do harder math problems by drawing out the story.

Name: _____   Date: _____   Time: ___:___

Problem of the Day #28: Calorie Count

Carmen is on a diet. She uses a calorie counting book to find the number of calories for the foods she eats each day. The local sandwich shop has 4 sandwiches that she likes. The number of calories are listed beside each sandwich.

| Sandwich Calories ||
Sandwich	Calories
Chicken Salad	515
Egg	280
Grilled Cheese	400
Ham	800
Seafood	460
Tuna	410

What is the Question?:
_____
_____
_____

Show Your Work:

Explain Your Thinking:

_____
_____
_____
_____

Bar Model/Tape Diagram, Base Ten Blocks, Number Grid, Open Number Line, Number Bond

Today's Affirmation: The more I talk about my thinking, the more connections I can make.

Name: _____ Date: _____ Time: ___:___

Problem of the Day #29: Walmart Sale

Melissa went to the Black Friday sale at Walmart. They had microwaves for $49, flat screen TV's for $139, bicycles for $79, and Nintendo's for $149. Melissa's mom gave her $75 dollars to spend. Her dad gave her $125 dollars to spend.

What is the Question?: _____

_____

Show Your Work:

Explain Your Thinking:

_____

_____

_____

_____

Bar Model/Tape Diagram, Base Ten Blocks, Number Grid, Open Number Line, Number Bond

Today's Affirmation: Everyone, I mean everyone makes mistakes, especially in math.

Name: _____   Date: _____   Time: ____:____

Problem of the Day #30: Collecting Box Tops

The students in grades K-5 at Jefferson Elementary School collected Box Tops to raise money for playground equipment. The table shows how many Box Tops each grade collected.

Box Tops	
Grade	Box Tops Collected
Kindergarten	515
First	320
Second	405
Third	625
Fourth	280
Fifth	400

What is the Question?:
_____
_____
_____

Show Your Work:

Explain Your Thinking:

_____
_____
_____
_____

Bar Model/Tape Diagram, Base Ten Blocks, Number Grid, Open Number Line, Number Bond

Today's Affirmation: I am smart in my own way and so is everyone else.

Name: _____   Date: _____   Time: ____:____

Problem of the Day #31: Treasure Box

There were 56 pencils in the teacher's treasure box. Caleb's mom sent in 30 more pencils for the treasure box. On Friday, 24 pencils were taken from the treasure box as prizes. How many pencils are left in the teacher's treasure box?

Show Your Work:

Explain Your Thinking:

_____
_____
_____
_____

Bar Model/Tape Diagram, Base Ten Blocks, Number Grid, Open Number Line, Number Bond, 100-Bead Number Line

Today's Affirmation: I can use anything as a math manipulative.

Name: _____   Date: _____   Time: ____:____

Problem of the Day #32: Jordan's Book

Jordan's favorite book has 198 pages. He read 116 pages during the week and 74 pages over the weekend. How many pages does he still need to read to finish his favorite book?

Show Your Work:

Explain Your Thinking:

_____

_____

_____

_____

Bar Model/Tape Diagram, Base Ten Blocks, Number Grid, Open Number Line, Number Bond, 100-Bead Number Line

Today's Affirmation: I don't give up!

Name: _____ Date: _____ Time: ___:____

Problem of the Day #33: The Parade

There were 246 people at the parade before it began. During the parade, 254 more people came. When it started raining, some people left. There are now 182 people at the parade. How many people left the parade when it started to rain?

Show Your Work:

Explain Your Thinking:

_____

_____

_____

_____

Bar Model/Tape Diagram, Base Ten Blocks, Number Grid, Open Number Line, 100-Bead Number Line

Today's Affirmation: My brain never stops growing with ideas.

Name: _____   Date: _____   Time: ____:____

Problem of the Day #34: The Fundraiser

Boy Scout Troop 17 sold tins of popcorn at the Farmer's Market last weekend. Their goal was to sell 500 tins. On Saturday, they sold 235 tins of popcorn. On Sunday, they sold 25 fewer tins of popcorn than they sold on Saturday. How many tins of popcorn do they still need to sell to reach their goal?

Show Your Work:

Explain Your Thinking:

_____

_____

_____

_____

Bar Model/Tape Diagram, Base Ten Blocks, Open Number Line, 100-Bead Number Line

Today's Affirmation: Math helps me build confidence in my abilities.

Name: _____ Date: _____ Time: ____:____

Problem of the Day #35: Chores

Nicky made $125 doing chores around the house this summer. She also made $75 babysitting her baby brother. When it was time for school, Nicky bought a new book bag for $29 which included tax. How much money does Nicky have left?

Show Your Work:

Explain Your Thinking:

_____

_____

_____

_____

Bar Model/Tape Diagram, Base Ten Blocks, Number Grid, Open Number Line, 100-Bead Number Line

Today's Affirmation: Every day that I keep trying, is one day closer to me getting it.

Name: _____ Date: _____ Time: ____:____

Problem of the Day #36: Lion King

Bonita's family of 4 traveled to New York to see "The Lion King" on Broadway. The family budgeted $700 for this trip. Each theater ticket costs $115 including tax. How much of what was budgeted remained after purchasing tickets to "The Lion King"?

Show Your Work:

Explain Your Thinking:

_____

_____

_____

_____

Bar Model/Tape Diagram, Base Ten Blocks, Number Grid, Open Number Line, 100-Bead Number Line, Equal Groups Model

Today's Affirmation: Believing that I can be successful in math means that I can do it!

Name: _____ Date: _____ Time: ____:____

Problem of the Day #37: BiC Gelocity Pens

My teacher bought a new pack of BiC Gelocity Pens for school. There were 24 pens in the pack. There were 4 red pens, 4 green pens, 4 blue pens, and the rest were black. How many black pens were in the pack?

Show Your Work:

Explain Your Thinking:

_____

_____

_____

_____

Bar Model/Tape Diagram, Base Ten Blocks, Number Line, Number Bond, Unifix/Snap Cubes, Equal Groups Model

Today's Affirmation: If I get lost, I will ask for help.

Name: _____ Date: _____ Time: ____:____

Problem of the Day #38: Colored Pencils

Malik's mom bought him a pack of 48 colored pencils for the beginning of school. On the first day, he gave 6 of the pencils to his friend and left some in his desk. When he returned home, he only had 18 pencils in his backpack. How many pencils did Malik leave in his desk?

Show Your Work:

Explain Your Thinking:

_____

_____

_____

_____

Bar Model/Tape Diagram, Base Ten Blocks, Open Number Line, Unifix/Snap Cubes

Today's Affirmation: Solving problems is what humans do!

Name: _____   Date: _____   Time: ____:____

Problem of the Day #39: Spring Break

While on spring break, Sarah visited Riverbanks Zoo, Ripley's Aquarium, and The African American Museum. She took twenty-four pictures at the zoo and sixty-six pictures at Ripley's Aquarium. She also took twenty pictures at The African American Museum.

What is the Question?: _____

_____

Show Your Work:

Explain Your Thinking:

_____

_____

_____

Bar Model/Tape Diagram, Base Ten Blocks, Number Line, Number Bond, Unifix/Snap Cubes, 100-Bead Number Line

Today's Affirmation: Not understanding something now, doesn't mean I won't in the future.

Name: _____   Date: _____   Time: ____:____

Problem of the Day #40: Alice's Favorite Hobby

Alice's favorite hobby is reading books. Over the weekend, she read 3 different books. The first book had 125 pages. The second book had 165 pages. If Alice read a total of 405 pages after reading all 3 books, how many pages are in the third book?

Show Your Work:

Explain Your Thinking:

_____

_____

_____

_____

Bar Model/Tape Diagram, Base Ten Blocks, Open Number Line, 100-Bead Number Line

Today's Affirmation: I love thinking! I am a thinker.

Name: _____ Date: _____ Time: ____:____

Problem of the Day #41: STEAM Camp

Each summer the state provides a free STEAM camp for third grade students. There are only 500 slots available for students. Jefferson Elementary School registered 198 students and Flora Elementary School registered 252 students. How many registration slots are still available for students to attend the STEAM camp?

Show Your Work:

Explain Your Thinking:

_____

_____

_____

_____

Bar Model/Tape Diagram, Base Ten Blocks, Open Number Line, 100-Bead Number Line

Today's Affirmation: Trying when I am nervous or scared shows I am brave and courageous!

Name: _____   Date: _____   Time: ____:____

Problem of the Day #42: McDonald's

McDonald's was open for Thanksgiving Day during the hours of 7 AM - 5 PM. A total of 605 people visited the restaurant that day. During breakfast, 71 people came in. During lunch, 254 people came in. After that, some more people came for dinner.

What is the Question?: _____

_____

Show Your Work:

Explain Your Thinking:

_____
_____
_____
_____

Bar Model/Tape Diagram, Base Ten Blocks, Open Number Line, 100-Bead Number Line

Today's Affirmation: I don't just memorize math facts! I understand math facts!

Name: _____ Date: _____ Time: ___:____

Problem of the Day #43: Apple Watch

Louis uses an Apple Watch to track the number of steps he takes at school each day. On Monday, he tracked 366 steps. On Tuesday, he tracked 420 steps. Wednesday was an early release day, so Louis tracked 208 less steps than he had on Tuesday. How many steps did Louis track this week so far?

Show Your Work:

Explain Your Thinking:

_____

_____

_____

_____

Bar Model/Tape Diagram, Base Ten Blocks, Open Number Line, 100-Bead Number Line

Today's Affirmation: I learn from the ideas of other people, even if I don't agree with them.

Name: _____   Date: _____   Time: ___:____

Problem of the Day #44: Beehive

There were about 1,000 bees in the beehive behind the farmer's house. This morning, about 250 bees left the beehive to collect pollen. About how many bees remained in the beehive?

Show Your Work:

Explain Your Thinking:

_____
_____
_____
_____

Bar Model/Tape Diagram, Base Ten Blocks, Open Number Line, 100-Bead Number Line

Today's Affirmation: I encourage myself and others by saying kind words.

Name: _____ Date: _____ Time: ____:____

Problem of the Day #45: Dave & Buster's

At Dave & Buster's, Henry won four hundred twenty-five tickets playing games. He spent one hundred eighty-five tickets on a hat. Later, he won one hundred thirty more tickets before going home. How many tickets did Henry take home with him?

Show Your Work:

Explain Your Thinking:

_____

_____

_____

_____

Bar Model/Tape Diagram, Base Ten Blocks, Open Number Line, 100-Bead Number Line

Today's Affirmation: I can learn something new every day.

Name: _____ Date: _____ Time: ___:___

Problem of the Day #46: Little Debbie Cakes

Shawn is buying Little Debbie snacks for his soccer team. He bought 5 boxes of fudge cakes. Each box has 5 fudge cakes in it. How many fudge cakes did Shawn buy?

Show Your Work:

Explain Your Thinking:

_____

_____

_____

_____

Square Tiles, Two-colored Counters, Unifix/Snap Cubes, Ten Frames, Equal Groups

Today's Affirmation: By helping others, I help myself learn more.

Name: _____ Date: _____ Time: ____:____

Problem of the Day #47: World's Largest Edible Mushroom

Fun Fact: The Termitomyces titanicus mushroom is the world's largest edible mushroom, with the cap capable of measuring a little more than three feet across. If there are 9 Termitomyces titanicus mushrooms growing wild, how many feet long could the caps measure altogether?

Show Your Work:

Explain Your Thinking:

_____

_____

_____

_____

Square Tiles, Two-colored Counters, Unifix/Snap Cubes, Yardstick, Open Number Line

Today's Affirmation: I like thinking about harder math because it helps me become smarter.

Name: _____   Date: _____   Time: ____:____

Problem of the Day #48: Agree or Disagree

Zuri looks at the expression 5 x 3 and says that another way to write this is 3 + 3 + 3 + 3 + 3. Do you agree or disagree with Zuri? Why?

Show Your Work:

Explain Your Thinking:

_____

_____

_____

_____

Square Tiles, Two-colored Counters, Unifix/Snap Cubes, Equal Groups

Today's Affirmation: Today is another opportunity to help someone else learn.

Name: _____   Date: _____   Time: ____:____

Problem of the Day #49: Pencils

Caleb's mom bought him 4 packs of pencils for school. There were 6 pencils in each pack. How many pencils does he have?

Show Your Work:

Explain Your Thinking:

_____

_____

_____

_____

Square Tiles, Two-colored Counters, Unifix/Snap Cubes, Equal Groups

Today's Affirmation: What matters is that I tried, even if I was the last one to finish.

Name: _____   Date: _____   Time: ____:____

Problem of the Day #50: Flower Delivery

On Valentine's Day, the florist delivered 5 vases of flowers to the school. Each vase had 10 flowers in it. How many flowers were delivered to the school?

Show Your Work:

Explain Your Thinking:

_____

_____

_____

_____

Square Tiles, Two-colored Counters, Unifix/Snap Cubes, Equal Groups

Today's Affirmation: I should share my thoughts because it might help someone else.

Name: _____ Date: _____ Time: ____:____

Problem of the Day #51: Bookshelves

Aaliyah has a bookshelf with 4 shelves. Each shelf has 5 books on it. Aaliyah's friend Bessie has a bookshelf with 6 shelves. Each shelf has 3 books on it.

What is the Question?: _____

_____

Show Your Work:

Explain Your Thinking:

_____

_____

_____

_____

Square Tiles, Two-colored Counters, Unifix/Snap Cubes

Today's Affirmation: I can do harder math problems by using math manipulatives.

Name: _____ Date: _____ Time: ____:____

Problem of the Day #52: Cars, Cars, Cars

A parking lot has 6 rows of parking spots. Each row has room for 6 cars. What is the maximum number of cars that can park in the parking lot?

Show Your Work:

Explain Your Thinking:

_____

_____

_____

_____

Square Tiles, Two-colored Counters, Unifix/Snap Cubes

Today's Affirmation: I am a proficient problem solver!

Name: _____ Date: _____ Time: ____:____

Problem of the Day #53: Lining Up

It was time for Liam's third grade class to go to lunch. The teacher had the students line up in 6 equal rows. There are 5 students in each row. How many students are in Liam's class?

Show Your Work:

Explain Your Thinking:

_____

_____

_____

_____

Square Tiles, Two-colored Counters, Unifix/Snap Cubes

Today's Affirmation: My ideas are worthy of sharing.

Name: _____ Date: _____ Time: ____:____

Problem of the Day #54: Kelly's Secret Number

Kelly has a secret number.
- ❏ When he doubles his number, he gets 16.
- ❏ When he triples the number, he gets 24.
- ❏ There are four sets of 2 in his number.

What number is Kelly's secret number?

Show Your Work:

Explain Your Thinking:

_____

_____

_____

_____

Square Tiles, Two-colored Counters, Unifix/Snap Cubes

Today's Affirmation: Everything learned was something we did not know at some point.

Name: _____ Date: _____ Time: ___:___

Problem of the Day #55: Traveling Ants

Some ants were going on a picnic. They were traveling eight in a row. There were six rows of ants. How many ants were going on the picnic?

Show Your Work:

Explain Your Thinking:

_____

_____

_____

_____

Square Tiles, Two-colored Counters, Unifix/Snap Cubes, Number Line, Ratio Table, Meter Stick

Today's Affirmation: Every day is a new opportunity to learn something new in math!

Name: _____ Date: _____ Time: ____:____

Problem of the Day #56: Tennis Balls

There are 8 cans of tennis balls in the gym. Each can has 3 tennis balls. How many tennis balls are there in all?

Show Your Work:

Explain Your Thinking:

_____

_____

_____

_____

Square Tiles, Two-colored Counters, Unifix/Snap Cubes, Number Line, Ratio Table, Meter Stick, Area/Array Model

Today's Affirmation: Forgetting and making mistakes are a part of learning and remembering.

Name: _____ Date: _____ Time: ____:____

Problem of the Day #57: Target

Mia's mom purchased 8 packs of Ivory Soap while shopping at Target. There were 8 bars of Ivory Soap in each pack. How many bars of soap did Mia's mom purchase?

Show Your Work:

Explain Your Thinking:

_____

_____

_____

_____

Square Tiles, Ten Frame, Two-colored Counters, Unifix/Snap Cubes, Number Line, Ratio Table, Equal Groups Model

Today's Affirmation: When I reason about math I get smarter.

Name: _____   Date: _____   Time: ____:____

Problem of the Day #58: Sharpening Pencils

Six girls were standing in line to sharpen their pencils. Each girl sharpened three pencils. How many pencils were sharpened?

Show Your Work:

Explain Your Thinking:

_____

_____

_____

_____

Square Tiles, Two-colored Counters, Unifix/Snap Cubes, Number Line, Ratio Table, Meter Stick, Equal Groups Model

Today's Affirmation: I am math. From the cells in my body to the hairs on my head. I am math.

Name: _____ Date: _____ Time: ____:____

Problem of the Day #59: Friendship Bracelets

Jada made friendship bracelets for her 8 friends. Each friendship bracelet had 9 beads. Cara also made friendship bracelets for her 4 friends. Each of Cara's friendship bracelet had 9 beads. How many more beads did Jada use than Cara to make her friendship bracelets?

Show Your Work:

Explain Your Thinking:

_____

_____

_____

_____

Square Tiles, Two-colored Counters, Unifix/Snap Cubes, Number Line, Ratio Table, Meter Stick, Equal Groups Model

Today's Affirmation: When I learn something new, I will forget some of it, and that is okay.

Name: _____ Date: _____ Time: ____:____

Problem of the Day #60: Pumpkin Seeds

The farmer planted 8 rows of pumpkin seeds in his garden. There were 7 seeds in each row. The neighboring farmer planted 4 rows of pumpkin seeds in his garden. They also had with 7 seeds in each row. How many pumpkin seeds did both farmers plant as a whole?

Show Your Work:

Explain Your Thinking:

_____

_____

_____

_____

Square Tiles, Two-colored Counters, Unifix/Snap Cubes, Number Line, Ratio Table, Meter Stick, Equal Groups Model

Today's Affirmation: I am a mathematical thinker.

Name: _____   Date: _____   Time: ____:____

Problem of the Day #61: Class Garden

Miranda's class planted a garden. They planted 4 rows of tomato plants. Each row had 10 tomatoes. They planted 4 rows of lettuce. Each row had 6 lettuces. They also planted 2 rows of squash. Each row has 4 squashes.

What is the Question?: _____

_____

Show Your Work:

Explain Your Thinking:

_____

_____

_____

_____

Square Tiles, Two-colored Counters, Unifix/Snap Cubes, Number Line, Ratio Table, Meter Stick, Area/Array Model

Today's Affirmation: I am brave and I take risks in math.

Name: _____ Date: _____ Time: ____:____

Problem of the Day #62: School Carnival

There was an apple dunking contest at the school carnival. George's teacher had 28 apples. She arranged the apples into 4 equal rows. How many apples are in each row?

Show Your Work:

Explain Your Thinking:

_____

_____

_____

_____

Square Tiles, Two-colored Counters, Unifix/Snap Cubes, Number Line, Ratio Table, Meter Stick, Area/Array Model

Today's Affirmation: Math is more than just getting an answer.

Name: _____ Date: _____ Time: ____:____

Problem of the Day #63: Bottles of Water

Mae's class collected plastic bottles of water to recycle. Mae's teacher asked her to line up 36 bottles of water in equal rows. What are 2 ways that Mae cold have lined up the water bottles?

Show Your Work:

Explain Your Thinking:

_____

_____

_____

_____

Square Tiles, Two-colored Counters, Unifix/Snap Cubes, Number Line, Ratio Table, Meter Stick, Area/Array Model

Today's Affirmation: When I do math, I feel like a superhero.

Name: _____   Date: _____   Time: ____:____

Problem of the Day #64: Bags of Apples

Sidney had 3 bags of apples. Her grandmother gave her 2 more bags of apples. Each bag had the same number of apples in them. If the bags had more than 4 apples, but less than 8 apples in it, how many apples could Sidney have?

Show Your Work:

Explain Your Thinking:

_____
_____
_____
_____

Square Tiles, Two-colored Counters, Unifix/Snap Cubes, Number Line, Ratio Table, Meter Stick, Equal Groups Model

Today's Affirmation: It is important that I explain and justify my thinking.

Name: _____ Date: _____ Time: ____:____

Problem of the Day #65: State Fair

At the State Fair, Bryan rode the bumper cars 2 times and the twister 4 times. Each ride costs 3 tickets. How many tickets did Bryan use for those rides?

Show Your Work:

Explain Your Thinking:

_____

_____

_____

_____

Square Tiles, Two-colored Counters, Unifix/Snap Cubes, Number Line, Ratio Table, Meter Stick, Equal Groups Model

Today's Affirmation: We are all smart in different ways.

Name: _____ Date: _____ Time: ____:____

Problem of the Day #66: Chewing Gum

Amira went to the bodega right after school on Tuesday to get snacks. She purchased 3 packs of chewing gum. Each pack had 5 sticks of gum inside. She gave her brother 3 sticks of chewing gum. How many sticks of gum does Amira have now?

Show Your Work:

Explain Your Thinking:

_____

_____

_____

_____

Square Tiles, Two-colored Counters, Unifix/Snap Cubes, Number Line, Meter Stick, Equal Groups Model

Today's Affirmation: Math helps me make sense of my world.

Name: _____ Date: _____ Time: ____:____

Problem of the Day #67: The Baker

The baker baked 2 batches of sugar cookies. Each batch had 8 cookies in it. She also baked 6 chocolate chip cookies. How many cookies did she bake?

Show Your Work:

Explain Your Thinking:

_____

_____

_____

_____

Square Tiles, Two-colored Counters, Unifix/Snap Cubes, Number Line, Meter Stick, Equal Groups Model

Today's Affirmation: Acting out math problems helps me understand what I am doing.

Name: _____ Date: _____ Time: ____:____

Problem of the Day #68: Recycling

Joey and his friend Caleb each collected 6 stacks of newspapers to be recycled. Each stack had 4 newspapers in it. How many newspapers did they collect?

Show Your Work:

Explain Your Thinking:

_____
_____
_____
_____

Square Tiles, Two-colored Counters, Unifix/Snap Cubes, Number Line, Meter Stick, Equal Groups Model

Today's Affirmation: Sketching out my thinking helps me see a problem more clearly.

Name: _____   Date: _____   Time: ___:____

Problem of the Day #69: Math Problems

Olivia had 8 math problems to complete. She spent about 8 minutes on each problem and then she spent 2 minutes checking over her answers. How many minutes did it take Olivia to finish?

Show Your Work:

Explain Your Thinking:

_____
_____
_____
_____

Square Tiles, Two-colored Counters, Unifix/Snap Cubes, Number Line, Meter Stick, Ten Frame, Area/Array Model

Today's Affirmation: Making mistakes is how I learn new things.

Name: _____ Date: _____ Time: ____:____

Problem of the Day #70: Moving Day

Arthur's family is moving to another state. Arthur must pack up his toys before moving day. His mom gave him 9 boxes to put his toys in. He puts 9 toys in each box. His mom finds 11 more toys under his bed. How many toys will Arthur pack?

Show Your Work:

Explain Your Thinking:

_____

_____

_____

_____

Square Tiles, Two-colored Counters, Unifix/Snap Cubes, Number Line, Meter Stick, Equal Groups Model

Today's Affirmation: I can help others by asking questions.

Name: _____   Date: _____   Time: ____:____

Problem of the Day #71: Pokémon Cards

Kevin has 3 red boxes and 5 blue boxes with his Pokémon cards inside. Each box has 8 Pokémon cards. How many Pokémon cards does Kevin have?

Show Your Work:

Explain Your Thinking:

_____

_____

_____

_____

Square Tiles, Two-colored Counters, Unifix/Snap Cubes, Number Line, Meter Stick, Equal Groups Model

Today's Affirmation: I can do harder math problems by acting out the story.

Name: _____ Date: _____ Time: ____:____

Problem of the Day #72: Chores

Last week, Anne cleaned the house for 5 hours and earned $5 for each hour. Then she got $10 for raking the leaves. How much money did she earn?

Show Your Work:

Explain Your Thinking:

_____

_____

_____

_____

Square Tiles, Two-colored Counters, Unifix/Snap Cubes, Number Line, Equal Groups Model, Money, Number Bond

Today's Affirmation: I am a problem solver.

Name: _____ Date: _____ Time: ____:____

Problem of the Day #73: Sharing

Toni had 4 bags of candy that she received at a friend's birthday party. Each bag has 9 pieces of candy in it. Toni gives her brother one of the bags. Later on that evening, Toni's mother gives her another bag of candy exactly like the ones she had. How many pieces of candy does Toni have now?

Show Your Work:

Explain Your Thinking:

_____

_____

_____

_____

Square Tiles, Two-colored Counters, Unifix/Snap Cubes, Number Line, Meter Stick, Equal Groups Model

Today's Affirmation: If I messed up yesterday, today is a new day, and I can try again.

Name: _____ Date: _____ Time: ____:____

Problem of the Day #74: School's Track Team

Mel has 1 week left to get ready to try out for her school's track team. She runs 5 miles each day Monday through Friday. During the weekend, she runs 10 miles each day. If Mel compared how many miles she ran during the week than on the weekend, what would she determine?

Show Your Work:

Explain Your Thinking:

_____

_____

_____

_____

Square Tiles, Two-colored Counters, Unifix/Snap Cubes, Number Line, Meter Stick, Equal Groups Model, Number Bond

Today's Affirmation: Math helps me develop grit and perseverance.

Name: _____ Date: _____ Time: ____:____

Problem of the Day #75: Prize Box

Mrs. Jefferson had a Class Prize Box that she used to reward students. Mrs. Jefferson bought 6 packs of pencils to add to her prize box. There were 8 pencils in each pack. Mrs. Jefferson gave 5 of the pencils to students who made the Honor Roll. How many pencils did she add to the prize box?

Show Your Work:

Explain Your Thinking:

_____

_____

_____

_____

Square Tiles, Two-colored Counters, Unifix/Snap Cubes, Number Line, Meter Stick, Equal Groups Model

Today's Affirmation: I don't need to be fast at math. I need to understand and that takes time.

Name: _____   Date: _____   Time: ____:____

Problem of the Day #76: Cereal Bars

Mrs. Brown bought 3 boxes of cereal bars for her 3 kids to have for breakfast each morning this week. Each box has 5 cereal bars in it. On Monday, Mrs. Brown puts a cereal bar in each of her kids' lunch boxes. How many cereal bars does she have left?

Show Your Work:

Explain Your Thinking:

_____

_____

_____

_____

Square Tiles, Two-colored Counters, Unifix/Snap Cubes, Number Line, Meter Stick, Equal Groups Model

Today's Affirmation: Think positive thoughts. Negative thoughts don't help us learn and grow.

Name: _____ Date: _____ Time: ____:____

Problem of the Day #77: Chairs

There are 4 rows of chairs set up in the gym. There are 8 chairs in each row. At the back of the gym, there are 12 additional chairs. How many chairs are there in the gym?

Show Your Work:

Explain Your Thinking:

_____

_____

_____

_____

Square Tiles, Two-colored Counters, Unifix/Snap Cubes, Number Line, Meter Stick, Equal Groups Model

Today's Affirmation: Mathematics makes me smarter because it makes me think!

Name: _____   Date: _____   Time: ____:____

Problem of the Day #78: The Farmer

The farmer planted a vegetable garden during the spring. He planted 6 rows of cucumbers with 9 plants in each row. He also planted 1 row of squash with 6 plants in a row. How many plants did the farmer plant?

Show Your Work:

Explain Your Thinking:

_____

_____

_____

_____

Square Tiles, Two-colored Counters, Unifix/Snap Cubes, Number Line, Meter Stick, Area/Array Model

Today's Affirmation: I am human and humans are mathematical beings.

Name: _____ Date: _____ Time: ____:____

Problem of the Day #79: Geese Crossing the Road

On my way to work this morning a flock of geese caused a traffic jam. The traffic stopped so the geese could cross the street. Their there were 6 rows of geese with 4 geese in each row. All the geese made it to the other side where 8 more geese were waiting for them. How many geese are there in the flock?

Show Your Work:

Explain Your Thinking:

_____

_____

_____

_____

Square Tiles, Two-colored Counters, Unifix/Snap Cubes, Number Line, Meter Stick, Area/Array Model

Today's Affirmation: Even when we know the answer, it can be fun to try out a different way.

Name: _____ Date: _____ Time: ___:____

Problem of the Day #80: Pet Store

Herschel, the clerk at the pet store, had arranged bird cages in two equal rows. There are 9 bird cages in each row. There are 12 additional bird cages in the back of the store. The store manager wanted Herschel to do an inventory of the number of bird cages in the pet store. How could Herschel have used mathematics to quickly tell the store manager how many bird cages were in the pet store?

Show Your Work:

Explain Your Thinking:

_____

_____

_____

_____

Square Tiles, Two-colored Counters, Unifix/Snap Cubes, Number Line, Meter Stick, Area/Array Model

Today's Affirmation: I have a powerful mind!

Name: _____   Date: _____   Time: ____:____

Problem of the Day #81: Juice Boxes

Malcolm brought two different kinds of juice boxes to the class party. He arranged the fruit punch juice boxes into 4 rows with 5 boxes in each row. Then, he arranged the orange juice boxes into 2 rows with 4 boxes in each row.

What is the Question?: _____

_____

Show Your Work:

Explain Your Thinking:

_____

_____

_____

_____

Square Tiles, Two-colored Counters, Unifix/Snap Cubes, Number Line, Meter Stick, Area/Array Model

Today's Affirmation: I am brilliant, bright, and getting better every day!

Name: _____   Date: _____   Time: ____:____

Problem of the Day #82: Box of Chocolates

Rosa purchased a large box of chocolates to share with her family during the holidays. The box had 2 layers of chocolate pieces. Each layer had 6 rows with 7 pieces of chocolate in each row. Rosa ate 10 pieces of chocolate on the subway before getting home. How many pieces of chocolate did Rosa have when she arrived home?

Show Your Work:

Explain Your Thinking:

_____

_____

_____

_____

Square Tiles, Two-colored Counters, Unifix/Snap Cubes, Number Line, Meter Stick, Area/Array Model

Today's Affirmation: Answers are important, but questioning and solving are just as important.

Name: _____ Date: _____ Time: ___:____

Problem of the Day #83: Field Trip

Third graders from Carlton Elementary School took a field trip to the planetarium to learn more about astronomy and the night sky. The students were seated in equal rows during the show. There were 10 rows with 8 students in each row. There were also 5 teachers and 5 parents standing beside the wall. How many people went on the field trip?

Show Your Work:

Explain Your Thinking:

_____

_____

_____

_____

Square Tiles, Two-colored Counters, Unifix/Snap Cubes

Today's Affirmation: New solutions happen when we try new ways of doing things.

Name: _____ Date: _____ Time: ____:____

Problem of the Day #84: Stickers

Annie has 15 stickers. She puts an equal number of stickers on 3 pages. How many stickers does Annie put on each page?

Show Your Work:

Explain Your Thinking:

_____

_____

_____

_____

Square Tiles, Two-colored Counters, Unifix/Snap Cubes, Number Line, Area/Array Model

Today's Affirmation: By asking questions, I learn more and help others learn as well.

Name: _____ Date: _____ Time: ____:____

Problem of the Day #85: Science Experiment

Mr. Owens must rearrange his students' desks so they can work together on a science experiment. He has 24 students and wants to split them into either 2, 3, or 4 equal groups. He was not sure how many equal groups he should make. Pick a number of equal groups for Mr. Owens to split the class into and figure out how many students would be in each group.

Show Your Work:

Explain Your Thinking:

_____

_____

_____

_____

Square Tiles, Two-colored Counters, Unifix/Snap Cubes, Number Line, Equal Groups Model

Today's Affirmation: I love learning! I am a learner.

Name: _____ Date: _____ Time: ____:____

Problem of the Day #86: The Bookshelf

Sean's mom bought him a new bookshelf. He thinks he has either 32 or 64 books. He wants to put an equal number of books on each of the 4 shelves. Guess which amount of books Sean has and figure out how many books will be on each shelf if he has that amount of books?

Show Your Work:

Explain Your Thinking:

_____
_____
_____
_____

Square Tiles, Two-colored Counters, Unifix/Snap Cubes, Number Line, Area/Array Model

Today's Affirmation: Everyone can become better at math! Including ME!

Name: _____   Date: _____   Time: ____:____

Problem of the Day #87: Bags of Candy

Dakota has 18 pieces of candy. She is putting the same amount of candy into 2 bags or 3 bags. She is undecided about how many bags she should use.

- ❏ If she uses 2 bags, how many pieces of candy will she put in each bag?
- ❏ If she uses 3 bags, how many pieces of candy will she put in each bag?

Show Your Work:

Explain Your Thinking:

_____

_____

_____

_____

Square Tiles, Two-colored Counters, Unifix/Snap Cubes, Number Line, Equal Groups Model

Today's Affirmation: I can do harder math problems by drawing out the story.

Name: _____ Date: _____ Time: ____:____

Problem of the Day #88: Marching Band

Maple Elementary School in Maryland has an amazing marching band. They use 24 students in the marching band for small events like school assemblies. They use 48 students in the marching band for large events like football games and parades. They always line up in 6 equal rows.

❏ How many students would be in each row of the marching band for small events?
❏ How many students would be in each row of the marching band for large events?

Show Your Work:

Explain Your Thinking:

_____

_____

_____

_____

Square Tiles, Two-colored Counters, Unifix/Snap Cubes, Number Line, Area/Array Model

Today's Affirmation: The more I talk about my thinking, the more connections I can make.

Name: _____ Date: _____ Time: ___:___

Problem of the Day #89: Marbles

Carter has 40 marbles. He wants to put them into 8 equal groups. Ben also has 40 marbles. He wants to put then into 5 equal groups. How many marbles will be in each group for both boys?

Show Your Work:

Explain Your Thinking:

_____

_____

_____

_____

Square Tiles, Two-colored Counters, Unifix/Snap Cubes, Number Line, Area/Array Model, Equal Groups Model

Today's Affirmation: Everyone, I mean everyone makes mistakes, especially in math.

Name: _____ Date: _____ Time: ____:____

Problem of the Day #90: Dog Treats

The veterinarian's assistant is preparing lunch for the dogs at the shelter. He has 37 Milk-Bone treats that must be shared equally between 9 dogs. He decides to save 1 of the treats so that he can remember what they looked like when he goes to the supermarket to get more. How many treats will each dog get?

Show Your Work:

Explain Your Thinking:

_____

_____

_____

_____

Square Tiles, Two-colored Counters, Unifix/Snap Cubes, Number Line, Area/Array Model, Equal Groups Model

Today's Affirmation: I am smart in my own way and so is everyone else.

Name: _____   Date: _____   Time: ____:____

Problem of the Day #91: Sharing Cookies

Cindy's mom baked 36 cookies. Cindy and her 5 friends shared them. Each person got the same number of cookies. How many cookies did each person get?

Show Your Work:

Explain Your Thinking:

_____

_____

_____

_____

Square Tiles, Two-colored Counters, Unifix/Snap Cubes, Number Line, Area/Array Model, Equal Groups Model

Today's Affirmation: I can use anything as a math manipulative.

Name: _____ Date: _____ Time: ____:____

Problem of the Day #92: Bead Necklaces

Chelsea is making a bead necklace for each of her friends. She has 50 glass beads in one pack and 50 ceramic beads in another pack. Each necklace uses 10 beads. How many necklaces can Chelsea make?

Show Your Work:

Explain Your Thinking:

_____

_____

_____

_____

Square Tiles, Two-colored Counters, Unifix/Snap Cubes, Number Line, Area/Array Model, Equal Groups Model

Today's Affirmation: I don't give up!

Name: _____   Date: _____   Time: ____:____

Problem of the Day #93: Counting Pennies

Carl is counting the pennies from his piggy bank. He places the pennies into rows with 8 pennies in each row. If he has 72 pennies, how many rows does he make?

Show Your Work:

Explain Your Thinking:

_____

_____

_____

_____

Square Tiles, Two-colored Counters, Unifix/Snap Cubes, Number Line, Ratio Table, Area/Array Model

Today's Affirmation: My brain never stops growing with ideas.

Name: _____   Date: _____   Time: ____:____

Problem of the Day #94: New Pencils

Matt received 3 packs of pencils with 12 colorful pencils in each pack for the school year. He opened the packs and put them in equal rows with 4 pencils in each row. How many rows will Matt have?

Show Your Work:

Explain Your Thinking:

_____

_____

_____

_____

Square Tiles, Two-colored Counters, Unifix/Snap Cubes, Number Line, Area/Array Model, Equal Groups Model

Today's Affirmation: Math helps me build confidence in my abilities.

Name: _____   Date: _____   Time: ____:____

Problem of the Day #95: Arranging Stickers

Christine received a pack of 28 large mermaid stickers for her birthday. She also received 14 small dolphin stickers. She arranges all the stickers into equal rows. She has 7 rows of stickers. How many stickers are in each row?

Show Your Work:

Explain Your Thinking:

_____

_____

_____

_____

Square Tiles, Two-colored Counters, Unifix/Snap Cubes, Number Line, Area/Array Model

Today's Affirmation: Every day that I keep trying, is one day closer to me getting it.

Name: _____ Date: _____ Time: ___:___

Problem of the Day #96: Fundraiser

My sister's class is sponsoring a fundraiser to raise money for new playground equipment. She must sell seventy-two chocolate bars to win a prize. If each box contains eight chocolate bars, how many boxes does she need to sell?

Show Your Work:

Explain Your Thinking:

_____

_____

_____

_____

Square Tiles, Two-colored Counters, Unifix/Snap Cubes

Today's Affirmation: Believing that I can be successful in math means that I can do it!

Name: _____ Date: _____ Time: ____:____

Problem of the Day #97: The Librarian

Our school librarian, Mrs. Weston, gave away 45 old library books to her student helpers. They each received the same number of books. She has 9 helpers. How many books did each helper get?

Show Your Work:

Explain Your Thinking:

_____

_____

_____

_____

Square Tiles, Two-colored Counters, Unifix/Snap Cubes, Number Line, Equal Groups Model, 100-Bead Rekenrek

Today's Affirmation: If I get lost, I will ask for help.

Name: _____ Date: _____ Time: ____:____

Problem of the Day #98: Friday Treat

As a treat for our math class last Friday, Mr. Thomas ordered donuts from Dunkin Donuts. He ordered 48 donuts. Dunkin Donuts can pack 6 or 12 donuts to a box.

❑ If Mr. Thomas chose 6 donuts to a box, how many boxes would Mr. Thomas get?
❑ If Mr. Thomas chose 12 donuts to a box, how many boxes would Mr. Thomas get?

Show Your Work:

Explain Your Thinking:

_____

_____

_____

_____

Square Tiles, Two-colored Counters, Unifix/Snap Cubes, Ten Frame, 100-Bead Rekenrek

Today's Affirmation: Solving problems is what humans do!

Name: _____ Date: _____ Time: ___:____

Problem of the Day #99: New Game

Justin and Adrianna are teaching their friends how to play a new game with marbles. Each person will need 3 marbles to play the game. Justine has the small jar that has 15 marbles. Adrianna has the large jar that has double the number of marbles as the small jar.

What is the Question?: _____

_____

Show Your Work:

Explain Your Thinking:

_____

_____

_____

_____

Square Tiles, Two-colored Counters, Unifix/Snap Cubes

Today's Affirmation: Not understanding something now, doesn't mean I won't in the future.

Name: _____   Date: _____   Time: ____:____

Problem of the Day #100: Square Tiles

Robert has 42 square tiles. If he places 6 square tiles in each row, how many rows of square tiles will there be?

Show Your Work:

Explain Your Thinking:

_____
_____
_____
_____

Square Tiles, Two-colored Counters, Unifix/Snap Cubes, Ten Frame, Meter Stick, Unit Cubes

Today's Affirmation: I love thinking! I am a thinker.

Name: _____ Date: _____ Time: ___:___

Problem of the Day #101: Maya's Rock Collection

This year Maya has 81 rocks in her collection. She placed them all on a table so her classmates could see them. She arranged 9 rocks in each row. Last year Maya had 63 rocks in her collection. She arranged them the same way for her classmates to see. How many more rows of rocks did Maya have this year than last year?

Show Your Work:

Explain Your Thinking:

_____
_____
_____
_____

Square Tiles, Two-colored Counters, Unifix/Snap Cubes, Ten Frame, 100-Bead Rekenrek

Today's Affirmation: Trying when I am nervous or scared shows I am brave and courageous!

Name: _____ Date: _____ Time: ____:____

Problem of the Day #102: Marching Band

Fayetteville State Marching Band performed during halftime at the football game. Seventy-two band members marched out onto the field in nine rows. How many band members were in each row?

Show Your Work:

Explain Your Thinking:

_____

_____

_____

_____

Square Tiles, Two-colored Counters, Unifix/Snap Cubes

Today's Affirmation: I don't just memorize math facts! I understand math facts!

Name: _____   Date: _____   Time: ____:____

Problem of the Day #103: Alex's Toy Cars

Alex wanted to sort 80 toy cars into 4 groups with the same number of toy cars in each group. He didn't want to place 1 car in each group one at a time. He felt that it would take too long. He used another strategy to get the toy cars into the 4 groups more quickly.

- ❏ What strategy could Alex have used?
- ❏ When he finished sorting the cars, how many cars would have be in each group?

Show Your Work:

Explain Your Thinking:

_____

_____

_____

_____

Square Tiles, Two-colored Counters, Unifix/Snap Cubes, Ten Frame, 100-Bead Rekenrek

Today's Affirmation: I learn from the ideas of other people, even if I don't agree with them.

Name: _____ Date: _____ Time: ____:____

Problem of the Day #104: Paper Butterflies

Lester is taping paper butterflies inside butterfly nets to hang on the wall in his bedroom. He has 27 paper butterflies and 3 nets. He wants to put the same number of paper butterflies in each net. How many paper butterflies will there be in each net?

Show Your Work:

Explain Your Thinking:

_____

_____

_____

_____

Square Tiles, Two-colored Counters, Unifix/Snap Cubes, Ten Frame, 100-Bead Rekenrek

Today's Affirmation: I encourage myself and others by saying kind words.

Name: _____   Date: _____   Time: ____:____

Problem of the Day #105: Winter Musical

Mr. Daniels is setting up the stage for the Winter Musical. He has to set up 40 chairs for the performance in equal rows. Draw a model to represent at least 2 ways that Mr. Daniels could set up the chairs.

Show Your Work:

Explain Your Thinking:

_____

_____

_____

_____

Square Tiles, Two-colored Counters, Unifix/Snap Cubes, Area/Array Model

Today's Affirmation: I can learn something new every day.

Name: _____   Date: _____   Time: ____:____

Problem of the Day #106: My Story

George was asked to write a story problem that can be represented by the expression 32 ÷ 4. What is an example of a story problem George could write?

Show Your Work:

Explain Your Thinking:

_____

_____

_____

_____

Square Tiles, Two-colored Counters, Unifix/Snap Cubes

Today's Affirmation: By helping others, I help myself learn more.

Name: _____ Date: _____ Time: ____:____

Problem of the Day #107: Gumballs

Frederick has 48 gumballs. He gives the gumballs to 8 friends so that each friend gets the same number. How many gumballs does each friend get?

Show Your Work:

Explain Your Thinking:

_____

_____

_____

_____

Square Tiles, Two-colored Counters, Unifix/Snap Cubes, Ten Frame, 100-Bead Rekenrek

Today's Affirmation: I like thinking about harder math because it helps me become smarter.

Name: _____ Date: _____ Time: ____:____

Problem of the Day #108: Barbie Sticker Album

Nevaeh received a sticker album and 40 Barbie stickers for a birthday gift. She saves 8 stickers. With the rest she places an equal number of the Barbie stickers on each of the 8 pages. How many Barbie stickers does Nevaeh put on each page?

Show Your Work:

Explain Your Thinking:

_____

_____

_____

_____

Square Tiles, Two-colored Counters, Unifix/Snap Cubes, Ten Frame, 100-Bead Rekenrek, Bar Model/Tape Diagram

Today's Affirmation: Today is another opportunity to help someone else learn.

Name: _____ Date: _____ Time: ____:____

Problem of the Day #109: Sleepover

Debra bought treats for her friends' sleepover. She bought 21 lollipops and 7 Blow Pops. Debra is going to share the treats equally with her 4 friends. She puts all of the treats into one bag and tells the friends to pick out a specific number of treats. How many treats will she tell each friend to pick so that each friend will get the same amount?

Show Your Work:

Explain Your Thinking:

_____

_____

_____

_____

Square Tiles, Two-colored Counters, Unifix/Snap Cubes, Ten Frame, 100-Bead Rekenrek

Today's Affirmation: What matters is that I tried, even if I was the last one to finish.

Name: _____   Date: _____   Time: ____:____

Problem of the Day #110: Bake Sale

Sherri's mom baked treats for the school bake sale. She baked 28 chocolate chip cookies and 36 brownies. The treats will be sold in bags of 8. How many bags of treats will she make?

Show Your Work:

Explain Your Thinking:

_____

_____

_____

_____

Square Tiles, Two-colored Counters, Unifix/Snap Cubes, Ten Frame, 100-Bead Rekenrek

Today's Affirmation: I should share my thoughts because it might help someone else.

Name: _____ Date: _____ Time: ___:___

Problem of the Day #111: The Zookeeper

The zookeeper at Riverbanks Zoo is going on vacation. Before leaving on Monday, he left each monkey 24 bananas and 18 mangos. The fruit should last for 7 days for each monkey. The zookeeper wanted every monkey to have at least one of each kind of fruit each day. He also wanted every monkey to have the same amount of fruit each day. How much fruit would every monkey have each day and what combination of bananas and mangoes could they have?

Show Your Work:

Explain Your Thinking:

_____

_____

_____

_____

Square Tiles, Two-colored Counters, Unifix/Snap Cubes, Ten Frame

Today's Affirmation: I can do harder math problems by using math manipulatives.

Name: _____ Date: _____ Time: ____:____

Problem of the Day #112: A Game of Soccer

At my school's end of year party, the teachers and students played a game of soccer together. There were 25 teachers and 35 students. If they divided into 6 equal teams, how many players are on each team?

Show Your Work:

Explain Your Thinking:

_____

_____

_____

_____

Square Tiles, Two-colored Counters, Unifix/Snap Cubes, Ten Frame, 100-Bead Number Line

Today's Affirmation: I am a proficient problem solver!

Name: _____ Date: _____ Time: ____ : ____

Problem of the Day #113: The Science Museum

Mayelly's class went on a field trip to the Science Museum. There were 4 teachers, 12 parents, and 24 students. During the field trip, they were asked to split into 8 equal groups to visit an exhibit. How many people were in each group?

Show Your Work:

Explain Your Thinking:

_____

_____

_____

_____

Square Tiles, Two-colored Counters, Unifix/Snap Cubes, Ten Frame, 100-Bead Rekenrek

Today's Affirmation: My ideas are worthy of sharing.

Name: _____ Date: _____ Time: ____:____

Problem of the Day #114: Bags of Candy

On Friday, Janet passed out bags of candy to some friends in her class. She put 24 pieces of candy into 6 bags. Each bag had the same amount of candy. If Janet gave away 5 bags of candy, how many pieces of candy did she give away?

Show Your Work:

Explain Your Thinking:

_____

_____

_____

_____

Square Tiles, Two-colored Counters, Unifix/Snap Cubes, Ten Frame, 100-Bead Rekenrek

Today's Affirmation: Everything learned was something we did not know at some point.

Name: _____ Date: _____ Time: ____:____

Problem of the Day #115: Paper Clips

Mrs. Hudson has 24 paper clips in one box and 32 paper clips in another box. Mrs. Hudson separates all of the paper clips into 4 equal groups. How many paper clips are there in each group?

Show Your Work:

Explain Your Thinking:

_____

_____

_____

_____

Square Tiles, Two-colored Counters, Unifix/Snap Cubes, Ten Frame, Paper Clips

Name: _____ Date: _____ Time: ____:____

Problem of the Day #116: The Holidays

Ruby was trying to finish her two favorite books before the Thanksgiving holiday. She had 24 pages left to read in one book and 30 pages left to read in the other book. Ruby will read the same number of pages each day for 6 days. How many pages will Ruby need to read each day from each book to finish both books in 6 days?

Show Your Work:

Explain Your Thinking:

_____
_____
_____
_____

Square Tiles, Two-colored Counters, Unifix/Snap Cubes, Ten Frame

Today's Affirmation: Forgetting and making mistakes are a part of learning and remembering.

Name: _____   Date: _____   Time: ____:____

Problem of the Day #117: Hockey Team Line

A hockey league allows 18-year-olds, 19-year-olds, and 20-year-olds to try out for their professional team. There were 7 players and 1 goalie from each age group that made the team. Now the coach has to decide how many lines of players he can have. A line of players include 5 players plus 1 goalie that usually play together during games. How many lines can this team create?

Show Your Work:

Explain Your Thinking:

_____

_____

_____

_____

Square Tiles, Two-colored Counters, Unifix/Snap Cubes, Ten Frame

Today's Affirmation: When I reason about math I get smarter.

Name: _____ Date: _____ Time: ____:____

Problem of the Day #118: Omelets

A chef combines 32 brown eggs and 16 white eggs to make 8 omelets. Each omelet uses the same number of eggs. The omelets are meant to be shared with 2 people.

❏   How many eggs will he use for each omelet?

❏   About how many eggs will each person eat, if they share the omelet as the chef intends?

Show Your Work:

Explain Your Thinking:

_____

_____

_____

_____

Square Tiles, Two-colored Counters, Unifix/Snap Cubes, Ten Frame, 100-Bead Rekenrek, Egg Cartons

Today's Affirmation: I am math. From the cells in my body to the hairs on my head. I am math.

Name: _____   Date: _____   Time: ____:____

Problem of the Day #119: Math Team

Twenty-four students tried out for the school's math team. Eighteen of the students were not selected for the team. The students who were selected for the math team were placed into two groups. How many students were placed in each group?

Show Your Work:

Explain Your Thinking:

_____

_____

_____

_____

Square Tiles, Two-colored Counters, Unifix/Snap Cubes, Ten Frame, Number Line, Equal Groups Model

Today's Affirmation: When I learn something new, I will forget some of it, and that is okay.

Name: _____   Date: _____   Time: ____:____

Problem of the Day #120: Summer Job

James made $55 mowing lawns over the summer. He spent $28 dollars buying new blades for his lawn mower. He spent the rest of his money to buy used video games. How many used video games that cost $9 could James buy from GameStop with the rest of his money?

Show Your Work:

Explain Your Thinking:

_____

_____

_____

_____

Square Tiles, Two-colored Counters, Unifix/Snap Cubes, Ten Frame, 100-Bead Rekenrek, Money (bills)

Today's Affirmation: I am a mathematical thinker.

Name: _____ Date: _____ Time: ____:____

Problem of the Day #121: National Puppy Day Sale

For National Puppy Day on March 23rd Petco had 26 puppies for sale on Monday. They were selling the puppies for $100 each when they normally would cost between $300 to $500. By the next day they had sold 16 of the puppies. The remaining puppies were housed in pairs. How many pairs of puppies were there?

Show Your Work:

Explain Your Thinking:

_____

_____

_____

_____

Square Tiles, Two-colored Counters, Unifix/Snap Cubes, Ten Frame, 100-Bead Rekenrek

Today's Affirmation: I am brave and I take risks in math.

Name: _____ Date: _____ Time: ___:___

Problem of the Day #122: Fun Park

Ray bought one hundred twenty tickets at Frankie's Fun Park. He used forty-eight tickets playing putt-putt and decided to use the rest on rides. If each ride is nine tickets, how many rides can Ray go on?

Show Your Work:

Explain Your Thinking:

_____

_____

_____

_____

Square Tiles, Two-colored Counters, Unifix/Snap Cubes, Ten Frame, 100-Bead Rekenrek

Today's Affirmation: Math is more than just getting an answer.

Name: _____ Date: _____ Time: ____:____

Problem of the Day #123: Rolls of Tape

My mom needs to buy some rolls of tape to wrap presents during the holidays. The tape is sold in packages of 2. There are 4 packages in a box. My Mom bought 7 boxes. How many rolls of tape does my mom buy?

Show Your Work:

Explain Your Thinking:

_____

_____

_____

_____

Square Tiles, Two-colored Counters, Unifix/Snap Cubes, Ten Frame, Table

Today's Affirmation: When I do math, I feel like a superhero.

Name: _____ Date: _____ Time: ____:____

Problem of the Day #124: Pencil Cap Eraser Sale

Mr. Black saw that pencil cap erasers were on sale at Walmart, so he bought 6 packs. He divided the packs fairly amongst his 3 kids. There were 10 pencil cap erasers in each pack. How many pencils cap erasers did each kid get?

Show Your Work:

Explain Your Thinking:

_____

_____

_____

_____

Square Tiles, Two-colored Counters, Unifix/Snap Cubes, Ten Frame

Today's Affirmation: It is important that I explain and justify my thinking.

Name: _____ Date: _____ Time: ___:___

Problem of the Day #125: Family Reunion T-Shirts

Zuri made t-shirts for her extended family for their family reunion. She bought 5 different colors of t-shirts, and she made 8 t-shirts in each color. Then, she divided up the t-shirts evenly and shipped them to 10 different families.

What is the Question?: _____

_____

Show Your Work:

Explain Your Thinking:

_____

_____

_____

_____

Square Tiles, Two-colored Counters, Unifix/Snap Cubes, Ten Frame, Equal Groups Model

Today's Affirmation: We are all smart in different ways.

Name: _____ Date: _____ Time: ____:____

Problem of the Day #126: Friendship Necklace Sale

Ella makes friendship necklaces to sell at the County Fair. She makes 4 different kinds of necklaces, and she brings 9 necklaces of each kind. When she gets to the County Fair, she divides the necklaces evenly among 6 racks to display them.

What is the Question?: _____

_____

Show Your Work:

Explain Your Thinking:

_____
_____
_____
_____

Square Tiles, Two-colored Counters, Unifix/Snap Cubes, Ten Frame, Area/Array Model

Today's Affirmation: Math helps me make sense of my world.

Name: _____ Date: _____ Time: ____:____

Problem of the Day #127: Cupcakes

Corey and his mother are baking cupcakes in the kitchen. Corey has a cupcake pan that can make 12 cupcakes at a time. His mother reminds him that 12 cupcakes is the same as 1 dozen. He uses the pan to make 3 batches of cupcakes. Then, he splits the cupcakes equally between 4 plates. He is going to keep one plate at home and take the others to school.

What is the Question?: _____

_____

Show Your Work:

Explain Your Thinking:

_____

_____

_____

_____

Square Tiles, Two-colored Counters, Unifix/Snap Cubes, Ten Frame Equal Groups Model

Today's Affirmation: Acting out math problems helps me understand what I am doing.

Name: _____ Date: _____ Time: ___:___

Problem of the Day #128: Peaches Galore

Mr. Washington went to the market to sell peaches. He divided 9 boxes of peaches evenly to display amongst 3 tables. Each box held 5 pounds of peaches. A pound of peaches is about 3 to 4 peaches.

What is the Question?: _____

_____

Show Your Work:

Explain Your Thinking:

_____

_____

_____

_____

Square Tiles, Two-colored Counters, Unifix/Snap Cubes, Ten Frame

Name: _____ Date: _____ Time: ___:___

Problem of the Day #129: The Glass Pitcher

Benjamin visited a local farmers market on Sunday. He paid $20 for a glass pitcher and 3 pounds of strawberries. The price of strawberries is $4 per pound. What is the price of the glass pitcher?

Show Your Work:

Explain Your Thinking:

_____

_____

_____

_____

Square Tiles, Two-colored Counters, Unifix/Snap Cubes, Ten Frame, Equal Groups Model, Number Line, Money (bills)

Today's Affirmation: Making mistakes is how I learn new things.

Name: _____ Date: _____ Time: ____:____

Problem of the Day #130: Fishing Trip

Kerri and her friends went on a fishing trip. They caught 4 fish. Each fish weighed about 9 pounds. There was a 50-pound limit on the total amount of fish caught by a group. Could they have caught more fish? If so, why?

Show Your Work:

Explain Your Thinking:

_____

_____

_____

_____

Square Tiles, Two-colored Counters, Unifix/Snap Cubes, Ten Frame, Base Ten Blocks

Today's Affirmation: I can help others by asking questions.

Name: _____   Date: _____   Time: ____:____

Problem of the Day #131: Cross Stitching Pictures

Duke's mom works from home cross stitching small pictures of famous African Americans to sell at flea markets across the state. This week, she cross stitched 4 pictures a day for 3 days. Last week, she cross stitched 2 pictures a day for 7 days.

What is the Question?: _____

_____

Show Your Work:

Explain Your Thinking:

_____

_____

_____

_____

Square Tiles, Two-colored Counters, Unifix/Snap Cubes, Ten Frame, Number Line, Area/Array Model

Today's Affirmation: I can do harder math problems by acting out the story.

Name: _____ Date: _____ Time: ___:____

Problem of the Day #132: Halves or Not?

For math homework, Iman was asked to draw shapes and split them into halves. What shapes can Iman draw that shows them split into halves?

Show Your Work:

Explain Your Thinking:

_____

_____

_____

_____

Cuisenaire Rods, Fraction Bars, Fraction Circles, Pattern Blocks, Geo Board

Today's Affirmation: I am a problem solver.

Name: _____ Date: _____ Time: ____:____

Problem of the Day #133: Equal Parts

Max drew a square and partitioned it into 4 equal parts one way. Then he drew another square and the partitioned it into 4 equal parts another way. What would each part be called?

Show Your Work:

Explain Your Thinking:

_____

_____

_____

_____

Cuisenaire Rods, Fraction Bars, Fraction Circles, Pattern Blocks, Geo Board

Today's Affirmation: If I messed up yesterday, today is a new day, and I can try again.

Name: _____   Date: _____   Time: ____:____

Problem of the Day #134: Partitioning Rectangles

Alvin drew a rectangle. He partitioned the rectangle into 2 equal parts. Then, he partitioned the same rectangle into 4 equal parts. What happened to the size of each part of the rectangle?

Show Your Work:

Explain Your Thinking:

_____

_____

_____

_____

Cuisenaire Rods, Fraction Bars, Fraction Circles, Pattern Blocks, Geo Board

Today's Affirmation: Math helps me develop grit and perseverance.

Name: _____ Date: _____ Time: ____:____

Problem of the Day #135: Representing Fractional Amounts

Draw a rectangle and divide it into 8 equal parts. Shade in 1 part.
- ❏ What fraction of the figure is shaded in?
- ❏ What fraction of the of the figure in not shaded?

Show Your Work:

Explain Your Thinking:

_____

_____

_____

_____

Cuisenaire Rods, Fraction Bars, Fraction Circles, Pattern Blocks, Geo Board

Today's Affirmation: I don't need to be fast at math. I need to understand and that takes time.

Name: _____ Date: _____ Time: ____:____

Problem of the Day #136: Brownies with Nuts

Michelle was asked to draw a rectangular or square pan of brownies and divide it into 2 equal parts. Then, she was asked to put nuts on one of the parts.

- ❏ Draw what Michelle's picture could look like.
- ❏ What fraction of the brownies have nuts?
- ❏ What fraction of the brownies don't have nuts?

Show Your Work:

Explain Your Thinking:

_____

_____

_____

_____

Cuisenaire Rods, Fraction Bars, Fraction Circles, Pattern Blocks, Geo Board

Today's Affirmation: Think positive thoughts. Negative thoughts don't help us learn and grow.

Name: _____ Date: _____ Time: ___:___

Problem of the Day #137: Pepperoni Pizza

Serena's mom made individual small pizzas for dinner last night. She cut the pizzas into 3 equal pieces. She added pepperoni to 1 of the pieces. What fraction of each pizza has pepperoni on it?

Show Your Work:

Explain Your Thinking:

_____

_____

_____

_____

Cuisenaire Rods, Fraction Bars, Fraction Circles, Pattern Blocks, Geo Board (circular side)

Today's Affirmation: Mathematics makes me smarter because it makes me think!

Name: _____ Date: _____ Time: ___:___

Problem of the Day #138: Representing Thirds

Tyra is learning about models for representing fractions. Tyra's teacher asked her to represent the fraction ⅓ in two different ways using different models. How could she represent ⅓?

Show Your Work:

Explain Your Thinking:

_____

_____

_____

_____

Fraction Bars, Fraction Circles, Pattern Blocks, Two-colored Counters, Number Line, Bar Model/Tape Diagram

Name: _____ Date: _____ Time: ____:____

Problem of the Day #139: Dividing Paper

Teddi was doing an art project with paper. Teddi divided a sheet of paper into 6 equal parts. He then colored 1 of the parts in blue. He colored the rest red. What fraction of the paper is not shaded in blue?

Show Your Work:

Explain Your Thinking:

_____
_____
_____
_____

Cuisenaire Rods, Fraction Bars, Fraction Circles, Pattern Blocks, Geo Board

Today's Affirmation: Even when we know the answer, it can be fun to try out a different way.

Name: _____ Date: _____ Time: ___:___

Problem of the Day #140: Richard's Homework

For homework, Richard was asked to draw a picture of some friends using stick figures. He had to label his work to show that ¼ of the group liked basketball, while the rest like tennis. What picture could Richard draw?

Show Your Work:

Explain Your Thinking:

_____

_____

_____

_____

Cuisenaire Rods, Fraction Bars, Fraction Circles, Pattern Blocks, Geo Board, Number Line, Two-colored counters

Today's Affirmation: I have a powerful mind!

Name: _____ Date: _____ Time: ____:____

Problem of the Day #141: Peach Pie for Dessert

Mo'Nique bought a miniature peach pie and ate ½ of it after dinner. Mo'Nique's friend, Cicely, bought a large peach pie and ate ½ of hers after dinner too. Did the girls eat the same amount of peach pie?

Show Your Work:

Explain Your Thinking:

_____

_____

_____

_____

Cuisenaire Rods, Fraction Bars, Fraction Circles, Pattern Blocks, Geo Board, Bar Model/Tape Diagram

Today's Affirmation: I am brilliant, bright, and getting better every day!

Name: _____   Date: _____   Time: ____:____

Problem of the Day #142: Completing Homework

Joe completed $\frac{4}{8}$ of his homework. Billy completed $\frac{8}{8}$ of his homework. Who completed more of their homework?

Show Your Work:

Explain Your Thinking:

_____

_____

_____

_____

Cuisenaire Rods, Fraction Bars, Fraction Circles, Pattern Blocks, Geo Board, Number Line, Bar Model/Tape Diagram

Today's Affirmation: Answers are important, but questioning and solving are just as important.

Name: _____ Date: _____ Time: ____:____

Problem of the Day #143: Catching the School Bus

Tina walks ½ of a mile to catch the school bus, John walks ¼ of a mile each day to catch the school bus, and Olivia walks ⅛ of a mile each day to catch the school bus.

What is the Question?: _____

_____

Show Your Work:

Explain Your Thinking:

_____

_____

_____

_____

Cuisenaire Rods, Fraction Bars, Fraction Circles, Pattern Blocks, Geo Board, Bar Model/Tape Diagram

Today's Affirmation: New solutions happen when we try new ways of doing things.

Name: _____ Date: _____ Time: ____:____

Problem of the Day #144: Which is Larger?

Emily was playing a game with Robbie. The game was called Figure IT Out! Emily gave Robbie directions:

- ❏ Draw 2 circles that are the same size.
- ❏ Divide the first circle into fourths and the second circle into halves.
- ❏ Lightly shade in ¼ of the first circle.
- ❏ Lightly shade in ½ of the second circle.

Emily then asked Robbie, which shaded portion represents the larger fraction? Robbie said that the ¼ was larger fraction. Was Robbie correct? Why or why not?

Show Your Work:

Explain Your Thinking:

_____

_____

_____

_____

Cuisenaire Rods, Fraction Bars, Fraction Circles, Pattern Blocks, Geo Board, Shapes Template

Today's Affirmation: By asking questions, I learn more and help others learn as well.

Name: _____   Date: _____   Time: ____:____

Problem of the Day #145: Which is Less?

Mrs. Brown baked two 9-inch pies. One was a pecan pie and the other was a pumpkin pie. The Brown family ate $3/6$ of the pecan pie and $5/6$ of the pumpkin pie after dinner. Which pie did the family eat the least of?

Show Your Work:

Explain Your Thinking:

_____

_____

_____

_____

Cuisenaire Rods, Fraction Bars, Fraction Circles, Pattern Blocks, Geo Board, Shapes Template

Today's Affirmation: I love learning! I am a learner.

Name: _____ Date: _____ Time: ___:___

Problem of the Day #146: New York City Public School's Population

In New York City Public Schools, $\frac{2}{4}$ of the total number of students are in elementary school, $\frac{1}{4}$ of the total number of students are in middle school, and $\frac{1}{4}$ of the total number of students are in high school.

What is the Question?: _____

_____

Show Your Work:

Explain Your Thinking:

_____

_____

_____

_____

Cuisenaire Rods, Fraction Bars, Fraction Circles, Pattern Blocks, Geo Board, Shapes Template

Today's Affirmation: Everyone can become better at math! Including ME!

Name: _____ Date: _____ Time: ____:____

Problem of the Day #147: Althea's Garden

Althea began planting vegetables in her garden during the spring. She planted carrots in $\frac{2}{6}$ of her garden, broccoli in $\frac{1}{6}$ of her garden, and spinach in $\frac{3}{6}$ of her garden.

What is the Question?: _____

_____

Show Your Work:

Explain Your Thinking:

_____

_____

_____

_____

Cuisenaire Rods, Fraction Bars, Fraction Circles, Pattern Blocks, Geo Board, Shapes Template, Bar Model/Tape Diagram

Today's Affirmation: I can do harder math problems by drawing out the story.

Name: _____ Date: _____ Time: ___:___

Problem of the Day #148: Movie Snacks

Alisha and Kim bought 2 king size Hershey bars to eat during the movie. Alisha ate $\frac{1}{6}$ of her candy bar and Kim ate $\frac{1}{3}$ of her candy bar. Who ate the most of their candy bar?

Show Your Work:

Explain Your Thinking:

_____

_____

_____

_____

Cuisenaire Rods, Fraction Bars, Fraction Circles, Pattern Blocks, Geo Board, Shapes Template

Today's Affirmation: The more I talk about my thinking, the more connections I can make.

Name: _____ Date: _____ Time: ____:____

Problem of the Day #149: Ordering Fractions

Calvin was given 4 different fractions and asked to put them in order from least to greatest. The fractions were ⅙, ½, ¼, and ⅛. How should Calvin order the fractions?

Show Your Work:

Explain Your Thinking:

_____

_____

_____

_____

Cuisenaire Rods, Fraction Bars, Fraction Circles, Pattern Blocks, Geo Board, Bar Model/Tape Diagram, Number Line

Today's Affirmation: Everyone, I mean everyone makes mistakes, especially in math.

Name: _____   Date: _____   Time: ____:____

Problem of the Day #150: Greater Than One-half

Clarence wanted to compete in the Fraction-A-Thon at his school. His teacher would only send the very best students who could explain and model their thinking. To see if he was ready for the competition, Clarence's teacher asked him to write and model a fraction that was greater than ½. What is a fraction that Clarence can write, model, and prove as greater than ½?

Show Your Work:

Explain Your Thinking:

_____

_____

_____

_____

Cuisenaire Rods, Fraction Bars, Fraction Circles, Pattern Blocks, Geo Board, Bar Model/Tape Diagram, Number Line

Today's Affirmation: I am smart in my own way and so is everyone else.

Name: _____ Date: _____ Time: ___:___

Problem of the Day #151: Equal to One Whole

Sheila was playing a game with her cousin. Sheila rolled a die. She rolled the number 1. Now she had to write, represent, and prove why the fraction she came up with was close to or equal to 1. What fraction could Sheila have come up with?

Show Your Work:

Explain Your Thinking:

_____

_____

_____

_____

Cuisenaire Rods, Fraction Bars, Fraction Circles, Bar Model/Tape Diagram, Number Line

Name: _____ Date: _____ Time: ___:___

Problem of the Day #152: Close to 0

Jay and Ivy were playing a fraction game during lunchtime. They had to pull a fraction card out of a bag to see whose fraction was closer to zero. Jay pulled out the fraction card of ¼. Ivy pulled the fraction card of ⅛. Which player pulled out a fraction card that is closer to 0?

Show Your Work:

Explain Your Thinking:

_____
_____
_____
_____

Cuisenaire Rods, Fraction Bars, Fraction Circles, Bar Model/Tape Diagram, Number Line

Today's Affirmation: I don't give up!

Name: _____   Date: _____   Time: ____:____

Problem of the Day #153: Going in Circles

Zyria and Toni were talking on the phone while they were creating a design on tote bags using circle halves. When they were finished, they texted each other how many halves they used.

- ❑ Zyria texted that she glued "$5/2$" on her tote bag.
- ❑ Toni texted that she glued "$8/2$" on her tote bag.

Toni said that she glued more halves on her tote bag. Zyria was not convinced and asked Toni to prove it. What could Toni have done to prove that she as correct?

Show Your Work:

Explain Your Thinking:

_____

_____

_____

_____

Cuisenaire Rods, Fraction Bars, Fraction Circles, Bar Model/Tape Diagram, Number Line

Today's Affirmation: My brain never stops growing with ideas.

Name: _____ Date: _____ Time: ____:____

Problem of the Day #154: Greater than 1

Omar was fascinated with fractions greater than 1. He asked Cora to compare the fractions 7/4 and 9/4 with him. Omar said that 7-fourths was less than 9-fourths and he wanted Cora to prove it using 2 different models and using the greater than, less than, or equal to symbols. Cora was up for the challenge. What could Cora have done to prove that Omar was correct or incorrect?

Show Your Work:

Explain Your Thinking:

_____

_____

_____

_____

Cuisenaire Rods, Fraction Bars, Fraction Circles, Bar Model/Tape Diagram, Number Line, Pattern Blocks

Today's Affirmation: Math helps me build confidence in my abilities.

Name: _____   Date: _____   Time: ___:____

Problem of the Day #155: Listing Fractions

Rick invented an app that listed fractions that matched a specific rule the user put in. He asked Naomi to try it out. She typed in, "List fractions greater than 1 using sixths as the denominator." What are some fractions the app could have listed?

Show Your Work:

Explain Your Thinking:

_____

_____

_____

_____

Cuisenaire Rods, Fraction Bars, Fraction Circles, Bar Model/Tape Diagram, Number Line, Pattern Blocks

Today's Affirmation: Every day that I keep trying, is one day closer to me getting it.

Name: _____   Date: _____   Time: ___:___

Problem of the Day #156: Center Time

Ray was partnered with Ke'Shawn for the Fraction's Center today. His first task was to draw and label fraction models that were equivalent to ½ and Ke'Shawn had to check to see if he was correct. What fractions could Ray have drawn and labeled for Ke'Shawn to check?

Show Your Work:

Explain Your Thinking:

_____

_____

_____

_____

Cuisenaire Rods, Fraction Bars, Fraction Circles, Bar Model/Tape Diagram, Number Line, Pattern Blocks

Today's Affirmation: Believing that I can be successful in math means that I can do it!

Name: _____ Date: _____ Time: ____:____

Problem of the Day #157: Which Fraction Doesn't Belong?

Danny was scrolling through Instagram and saw a "Which One Doesn't Belong" riddle from one of his favorite math influencers. Danny was asked to look at 4 fractional amounts written in unit form. He had to decide which one doesn't belong.

Which fraction could Danny say doesn't belong? Why?

2-fourths	6-fourths
3-sixths	4-eighths

Show Your Work:

Explain Your Thinking:

_____

_____

_____

_____

Cuisenaire Rods, Fraction Bars, Fraction Circles, Bar Model/Tape Diagram, Number Line, Pattern Blocks

Today's Affirmation: If I get lost, I will ask for help.

Name: _____ Date: _____ Time: ____:____

Problem of the Day #158: Subway Sandwich for Lunch

Arie ate ½ of a Subway sandwich for lunch. Her best friend Sonya ate ²⁄₄ of the same size Subway sandwich. Who ate more of their Subway sandwich?

Show Your Work:

Explain Your Thinking:

_____

_____

_____

_____

Cuisenaire Rods, Fraction Bars, Fraction Circles, Bar Model/Tape Diagram, Number Line

Today's Affirmation: Solving problems is what humans do!

Name: _____ Date: _____ Time: ____:____

Problem of the Day #159: Lines, Lines, Lines

Zuly drew 2 number lines on her paper that are the same length. She divided the first number line into halves and placed a point where ½ would be. She then divided the second number line into fourths and placed a point where 2/4 would be. What do you notice about the number lines Zuly drew?

Show Your Work:

Explain Your Thinking:

_____

_____

_____

_____

Cuisenaire Rods, Fraction Bars, Fraction Circles, Bar Model/Tape Diagram, Number Line

Today's Affirmation: Not understanding something now, doesn't mean I won't in the future.

Name: _____ Date: _____ Time: ___:___

Problem of the Day #160: Folding Paper

Mike was asked to fold a sheet of paper in half. He saw that he had 2 equal parts. He shaded one of the parts. Then, Mike was asked to fold the 2 halves in half again. When he unfolded the paper, he had 4 equal parts. Doing this helped Mike to compare these fractions. Make a model to show what Mike's folded paper looked like. Use this to compare ½ to ²⁄₄ using the symbols (>, <, or =).

Show Your Work:

Explain Your Thinking:

_____

_____

_____

_____

Cuisenaire Rods, Fraction Bars, Fraction Circles, Bar Model/Tape Diagram, Number Line, Pattern Blocks

Today's Affirmation: I love thinking! I am a thinker.

Name: _____ Date: _____ Time: ____:____

Problem of the Day #161: Justine's Thinking

Justine and Yuri were cutting some wooden planks. Justine cut off $\frac{4}{8}$ of his plank. Yuri cut off $\frac{1}{2}$ of his plank. Justine said that they cut off the same amount. Do you agree or disagree?

Show Your Work:

Explain Your Thinking:

_____

_____

_____

_____

Cuisenaire Rods, Fraction Bars, Fraction Circles, Bar Model/Tape Diagram, Number Line

Today's Affirmation: Trying when I am nervous or scared shows I am brave and courageous!

Name: _____   Date: _____   Time: ____:____

Problem of the Day #162: Torn Index Cards

Sharrai tore off two-thirds of an index card and gave it to Ally to write his email address on. Ally tore off four-sixths of another index card out of the same pack and wrote his number on it and gave it to Sharrai. Did Sharrai and Ally give each other the same sized portion of the index cards? Why or why not?

Show Your Work:

Explain Your Thinking:

_____

_____

_____

_____

Cuisenaire Rods, Fraction Bars, Fraction Circles, Bar Model/Tape Diagram, Number Line, Index Cards

Today's Affirmation: I don't just memorize math facts! I understand math facts!

Name: _____ Date: _____ Time: ____:____

Problem of the Day #163: Not Equivalent

Robert was asked by another 3rd grader to name 2 fractions that are not equivalent. What 2 fractions could Robert name and explain why these fractions are not equivalent.

Show Your Work:

Explain Your Thinking:

_____

_____

_____

_____

Cuisenaire Rods, Fraction Bars, Fraction Circles, Bar Model/Tape Diagram, Number Line, Pattern Blocks

Today's Affirmation: I learn from the ideas of other people, even if I don't agree with them.

Name: _____ Date: _____ Time: ____:____

Problem of the Day #164: Pie Eating Contest

Miranda ate ½ of a peach pie. Tonya ate ³⁄₆ of a peach pie that was the same size. Tonya said she ate more of her pie than Miranda because she ate more pieces. Do you agree or disagree?

Show Your Work:

Explain Your Thinking:

_____

_____

_____

_____

Cuisenaire Rods, Fraction Bars, Fraction Circles, Bar Model/Tape Diagram, Number Line

Today's Affirmation: I encourage myself and others by saying kind words.

Name: _____   Date: _____   Time: ____:____

Problem of the Day #165: Making Pizza

Mitch's class was asked to create a pizza model and represent their toppings using at least 2 equivalent fractions. Mitch cut his pizza into 8 equal slices. He put buffalo chicken on 2 of the slices. He put pineapples on 4 of the slices, and he put onions on the remaining 2 slices. How could Mitch describe his pizza using equivalent fractions?

Show Your Work:

Explain Your Thinking:

_____

_____

_____

_____

Cuisenaire Rods, Fraction Bars, Fraction Circles, Bar Model/Tape Diagram, Number Line, Shapes Template

Today's Affirmation: I can learn something new every day.

Name: _____ Date: _____ Time: ____:____

Problem of the Day #166: How Are They Similar?

Mr. Adams asked his students to list as many ways as possible to show that ½ and ²⁄₄ are similar. What are some responses that the students could have given?

Show Your Work:

Explain Your Thinking:

_____

_____

_____

_____

Cuisenaire Rods, Fraction Bars, Fraction Circles, Bar Model/Tape Diagram, Number Line, Shapes Template

Today's Affirmation: By helping others, I help myself learn more.

Name: _____ Date: _____ Time: ___:___

Problem of the Day #167: Yani's Pie

Yani baked a sweet potato pie for her family. She sliced the pie into 8 equal pieces. Her little brother ate 2 slices, her mom ate 1 slice, and her dad ate 4 slices.

What is the Question?: _____

_____

Show Your Work:

Explain Your Thinking:

_____

_____

_____

_____

Cuisenaire Rods, Fraction Bars, Fraction Circles, Bar Model/Tape Diagram, Number Line, Shapes Template

Today's Affirmation: I like thinking about harder math because it helps me become smarter.

Name: _____ Date: _____ Time: ____:____

Problem of the Day #168: Finding Equivalent Fractions

Debbie and Cici wanted to find fractions equivalent to ½. To show a fraction equivalent to ½, Debbie drew a rectangle and divided it into fourths. Cici drew the same size rectangle and divided it into eighths. Draw a picture to show what both Debbie and Cici drew. What fraction of each girl's picture is equivalent to ½?

Show Your Work:

Explain Your Thinking:

_____

_____

_____

_____

Cuisenaire Rods, Fraction Bars, Fraction Circles, Bar Model/Tape Diagram, Number Line, Shapes Template

Today's Affirmation: Today is another opportunity to help someone else learn.

Name: _____   Date: _____   Time: ____:____

Problem of the Day #169: Time to Finish Homework

Lori's math homework took her 35 minutes to complete. Her science homework took her 25 minutes. How much time did it take Lori to finish her math and science homework?

Show Your Work:

Explain Your Thinking:

_____

_____

_____

_____

Judy Clock

Today's Affirmation: What matters is that I tried, even if I was the last one to finish.

Name: _____   Date: _____   Time: ___:____

Problem of the Day #170: Math Class

Debbie's math class starts at 10:05 a.m. Her math class finishes at 10:50 a.m. How long is Debbie in math class?

Show Your Work:

Explain Your Thinking:

_____

_____

_____

_____

Judy Clock, Open Number Line

Today's Affirmation: I should share my thoughts because it might help someone else.

Name: _____   Date: _____   Time: ____:____

Problem of the Day #171: Movie Time

Angela was excited to see the latest Marvel Comic Movie. Angela arrives at the theater at 6:15 p.m. It takes her 20 minutes to buy tickets and 10 minutes to get to her seat. Her movie began 5 minutes later. What time did Angela's movie start?

Show Your Work:

Explain Your Thinking:

_____

_____

_____

_____

Judy Clock, Open Number Line

Today's Affirmation: I can do harder math problems by using math manipulatives.

Name: _____ Date: _____ Time: ___:___

Problem of the Day #172: Bedtime

Marcy reads a book every night before bed. She starts reading at 7:30 p.m. and finishes reading at 7:55 p.m. How long does Marcy read each night?

Show Your Work:

Explain Your Thinking:

_____
_____
_____
_____

Judy Clock, Open Number Line

Today's Affirmation: I am a proficient problem solver!

Name: _____ Date: _____ Time: ____:____

Problem of the Day #173: Wrestling Event

Alex's wrestling event started at 3:30 p.m. He left about an hour ago to get warmed up. Alex wanted to get to the event so he could see his first wrestling match of the day. If it took 25 minutes to get to the event location and 10 minutes to find parking, what time should Alex leave his house?

Show Your Work:

Explain Your Thinking:

_____

_____

_____

_____

Judy Clock, Open Number Line

Today's Affirmation: My ideas are worthy of sharing.

Name: _____ Date: _____ Time: ____:____

Problem of the Day #174: Basketball Practice

Sean played for the Cooper's Town Youth Basketball League. Sean's basketball practice lasts 45 minutes. If practice ends at 6:50 p.m., what times does practice begin?

Show Your Work:

Explain Your Thinking:

_____

_____

_____

_____

Judy Clock, Open Number Line, Number Grid

Today's Affirmation: Everything learned was something we did not know at some point.

Name: _____ Date: _____ Time: ____:____

Problem of the Day #175: Polygons

Kaitlin read the following riddle:
- ❏ We are regular polygons.
- ❏ We can also be classified as quadrilaterals.
- ❏ Our opposite sides are equal.

What polygons could we be?

Show Your Work:

Explain Your Thinking:

_____
_____
_____
_____

Pattern blocks, Shapes Template, Attribute Blocks, Geo Board

Today's Affirmation: Every day is a new opportunity to learn something new in math!

Name: _____ Date: _____ Time: ___:___

Problem of the Day #176: Describing Quadrilaterals

Kendra was asked to pull 2 quadrilaterals from a bag without looking and write 2 statements telling how they are alike and 2 statements telling how they are different. She pulled a square and a trapezoid out of the bag. What could she write about the two quadrilaterals?

Show Your Work:

Explain Your Thinking:

_____

_____

_____

_____

Pattern blocks, Shapes Template, Attribute Blocks, Geo Board

Today's Affirmation: Forgetting and making mistakes are a part of learning and remembering.

Name: _____   Date: _____   Time: ___:___

Problem of the Day #177: Guess My Quadrilateral

Aaron was playing a game called, "Guess My Quadrilateral" with his friend Chris. He gave Chris three clues about a quadrilateral that he drew. Chris had to guess the name of the quadrilateral. Aaron said his quadrilateral has 4 sides, 4 square corners or right angles, and the opposite sides are congruent. Which polygon(s) could Aaron draw?

Show Your Work:

Explain Your Thinking:

_____
_____
_____
_____

Pattern blocks, Shapes Template, Attribute Blocks, Geo Board

Today's Affirmation: When I reason about math I get smarter.

Name: _____ Date: _____ Time: ____:____

Problem of the Day #178: Making Polygons

Ciara was given a square and two small right triangles from a set of tangrams. She was asked to create a new polygon using those shapes. What polygon(s) could Ciara create? Draw the polygon(s) Ciara could have created. Try to identify the mathematical name the polygon(s) you draw.

Show Your Work:

Explain Your Thinking:

_____

_____

_____

_____

Tangrams, Pattern blocks, Shapes Template, Attribute Blocks, Geo Board, Ruler, Tangrams

Today's Affirmation: I am math. From the cells in my body to the hairs on my head. I am math.

Name: _____   Date: _____   Time: ____:_____

Problem of the Day #179: Sorting Quadrilaterals

Colin was given 5 quadrilaterals and asked to sort them into 2 different groups based on their attributes. He was given a rectangle, square, rhombus, parallelogram, and trapezoid. What are 2 groups he could sort the quadrilaterals into?

Show Your Work:

Explain Your Thinking:

_____

_____

_____

_____

Pattern blocks, Shapes Template, Attribute Blocks, Geo Board

Today's Affirmation: When I learn something new, I will forget some of it, and that is okay.

Name: _____   Date: _____   Time: ____:____

Problem of the Day #180: Draw It

During his math test, Emmett was asked to draw a picture of a quadrilateral and a picture of a non-quadrilateral. What could Emmett draw?

Show Your Work:

Explain Your Thinking:

_____

_____

_____

_____

Pattern blocks, Shapes Template, Attribute Blocks, Geo Board, Ruler

Today's Affirmation: _____

Name: _____ Date: _____ Time: ____:____

Problem of the Day #_____: _____

Show Your Work:

Explain Your Thinking:

_____

_____

_____

_____

Today's Affirmation: _____

Name: _____ Date: _____ Time: ____:____

Problem of the Day #_____: _____

Show Your Work:

Explain Your Thinking:

_____

_____

_____

_____

Today's Affirmation: _____

Name: _____ Date: _____ Time: ____:____

Problem of the Day #_____: _____

Show Your Work:

Explain Your Thinking:

_____

_____

_____

_____

Today's Affirmation: _____

Name: _____ Date: _____ Time: ____:____

Problem of the Day #_____: _____

Show Your Work:

Explain Your Thinking:

_____

_____

_____

_____

Today's Affirmation: _____

Name: _____ Date: _____ Time: ____:____

Problem of the Day #_____: _____

Show Your Work:

Explain Your Thinking:

_____

_____

_____

_____

Today's Affirmation: _____

Name: _____ Date: _____ Time: ____:____

Problem of the Day #_____: _____

Show Your Work:

Explain Your Thinking:

_____

_____

_____

_____

Today's Affirmation: _____

Name: _____ Date: _____ Time: ____:____

Problem of the Day #_____: _____

Show Your Work:

Explain Your Thinking:

_____

_____

_____

_____

Today's Affirmation: _____

Name: _____ Date: _____ Time: ____:____

Problem of the Day #_____: _____

Show Your Work:

Explain Your Thinking:

_____

_____

_____

_____

Today's Affirmation: _____

Name: _____ Date: _____ Time: ____:____

Problem of the Day #_____: _____

Show Your Work:

Explain Your Thinking:

_____

_____

_____

_____

Today's Affirmation: _____

Name: _____   Date: _____   Time: ____:____

Problem of the Day #_____: _____

Show Your Work:

Explain Your Thinking:

_____

_____

_____

_____

Today's Affirmation: _____

Name: _____ Date: _____ Time: ____:____

Problem of the Day #_____: _____

Show Your Work:

Explain Your Thinking:

_____

_____

_____

_____

Today's Affirmation: _____

Name: _____ Date: _____ Time: ____:____

Problem of the Day #_____: _____

Show Your Work:

Explain Your Thinking:

_____

_____

_____

_____

Today's Affirmation: _____

Name: _____ Date: _____ Time: ____:____

Problem of the Day #_____: _____

Show Your Work:

Explain Your Thinking:

_____

_____

_____

_____

Today's Affirmation: _____

Name: _____ Date: _____ Time: ____:____

Problem of the Day #_____: _____

Show Your Work:

Explain Your Thinking:

_____

_____

_____

_____

Today's Affirmation: _____

Name: _____ Date: _____ Time: ____:____

Problem of the Day #_____: _____

Show Your Work:

Explain Your Thinking:

_____

_____

_____

_____

Today's Affirmation: _____

Name: _____ Date: _____ Time: ____:____

Problem of the Day #_____: _____

Show Your Work:

Explain Your Thinking:

_____

_____

_____

_____

Today's Affirmation: _____

Name: _____ Date: _____ Time: ____:____

Problem of the Day #_____: _____

Show Your Work:

Explain Your Thinking:

_____

_____

_____

_____

Today's Affirmation: _____

Name: _____ Date: _____ Time: ____:____

Problem of the Day #_____: _____

Show Your Work:

Explain Your Thinking:

_____

_____

_____

_____

Today's Affirmation: _____

Name: _____ Date: _____ Time: ____:____

Problem of the Day #_____: _____

Show Your Work:

Explain Your Thinking:

_____

_____

_____

_____

Today's Affirmation: _____

Name: _____ Date: _____ Time: ____:____

Problem of the Day #_____: _____

Show Your Work:

Explain Your Thinking:

_____

_____

_____

_____

My Math Affirmations

What are some math affirmations that you want to remember? They could be from this book or ones you created. Remember, your thoughts and words have power. **You are powerful!**

_____

_____

_____

_____

_____

_____

_____

_____

_____

_____

_____

_____

_____

_____

3rd Grade Reflection

Take a moment to reflect on your progress. Write a letter to your teacher. Which of your goals did you achieve? Which of your goals are you still working towards? What are your next steps?

Date: _____

_____,

_____

_____

_____

_____

_____

_____

_____

_____

_____

_____

_____

Made in the USA
Middletown, DE
07 June 2025

76629866R00124